Between Forest and Sky

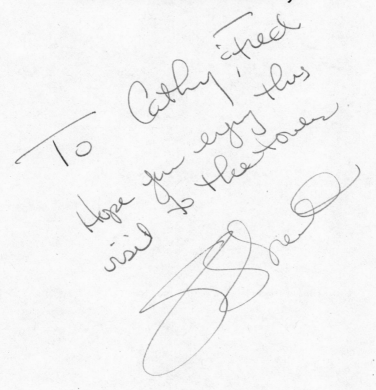

SHARON STRATTON

Between Forest and Sky
a fire-tower journal

Heritage
House

VICTORIA · VANCOUVER · CALGARY

Heritage House Publishing Company Ltd.
#108 – 17665 66A Avenue
Surrey, BC V3S 2A7
www.heritagehouse.ca

Heritage House Publishing Company Ltd.
PO Box 468
Custer, WA
98240-0468

Library and Archives Canada Cataloguing in Publication
Stratton, Sharon
    Between forest and sky: a fire-tower journal / Sharon Stratton.

ISBN-13: 978-1-894974-16-5
ISBN-10: 1-894974-16-6

    1. Stratton, Sharon. 2. Fire lookout stations—Alberta. 3. Women fire lookouts—Alberta—Biography. 4. Fire lookouts—Alberta—Biography.
I. Title.

SD421.375.S78   2006 363.37'9   C2006-904947-5

Library of Congress Control Number: 2006932766

Edited by Brian Bindon and Ursula Vaira
Proofread by Marial Shea
Book design by Carolyn Boyd
Cover design by Frances Hunter
Cover and interior photos by Sharon Stratton

Printed in Canada

Heritage House acknowledges the financial support for its publishing program from the Government of Canada through the Book Publishing Industry Development Program (BPIDP), Canada Council for the Arts, and the province of British Columbia through the British Columbia Arts Council and the Book Publishing Tax Credit.

The Canada Council | Le Conseil des Arts
    for the Arts | du Canada

BRITISH COLUMBIA
ARTS COUNCIL
Supported by the Province of British Columbia

This book has been produced on 100% post-consumer recycled paper, processed chlorine free and printed with vegetable-based dyes.

For my mother,
June Helena Victoria (Sward) Stratton,
who always had the faith and strength
to let me wander into the woods by myself

Contents

Introduction

Am I really that different from you? I don't believe so.

There are ... who can imagine ... perhaps millions of paths that each of us has a choice to take, if a number can be attributed to that thought at all. Mine just happened to go through the forest.

From the time I learned to climb the tree in the family backyard, participated in family hikes through the Niagara region's many parks, took weekend hacks along the Niagara Escarpment atop my horse and viewed the shorelines that drifted past as I kayaked, my path always brought me to the peace of the forest.

Camping in provincial parks became backpacking in Algonquin Park, followed by cross-country skiing, then kayaking the shores of Georgian Bay. No matter what method I used to gain access to the forest, once inside I would think of another perspective that I hadn't yet enjoyed and endeavour to experience that as well.

But don't imagine me a jock or woodsman—I was a city gal. As my career pursuits saw me through 5 years as a veterinary technician followed by 13 years with a daily newspaper, the forest was an escape or vacation destination.

While I was at university at the grand old age of around 38 years old, I leafed through a favourite magazine one day and stumbled upon a story about a woman who had won a quilting contest sponsored by that magazine. She would take her raw cotton cloth out to her fire tower, where she would concoct her own natural dyes and complete a quilt each season. My imagination soared! I hadn't realized there were still manned fire towers in Canada.

I couldn't get it from my mind. The romance of it all, being alone in the woods from the spring flush of leaves until the colours turned and fell to the forest floor. I had to do this! And so began my annual December tradition of applying to Alberta Sustainable Resources.

After five years of applications, it finally happened.

With my university degree finished, my wanderings had led me to Calgary, Alberta. One winter's day in early 2002 there was a message on my answering machine from the Jordan Forestry Office, asking if I would be interested in a tower position.

Oh, the squeals I squealed that day—the dances I danced! The friends I called, playing and replaying that message!

My dream had come true, and so began yet another metamorphosis of the ongoing relationship between the forest and me. After replying to the Jordan office, I was off to the Alberta Sustainable Resources Training Centre in Hinton for a week's studies. Only a few weeks after that, I was standing at the base of Connaught, my first tower, thinking about how daunting that 30-metre climb looked.

This book contains my particular adventures and misadventures, accomplishments and foibles. A good friend and seasoned towerman once told me that even after almost 20 years, he had yet to find the tie that binds the 130 or so towerpeople in the Alberta firetower program. I agree. For as many towerfolk as you could share stories with, I doubt you'd find two who look out their cupola windows and see things from the same perspective.

These are my personal stories.

While the adventures are all true, names of people, places, towers and lookouts have been changed out of respect for the privacy of my co-workers. If you should ever happen upon a fire tower in your travels, please endeavour to respect the towerperson's time, space and privacy.

— S.S.

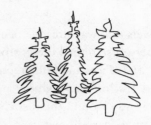

Connaught Tower: Little Cabin on the Muskeg

What an adventure reaching the Jordan Forestry Office turned out to be! I had crammed all my gear, my nine-year-old Irish wolfhound, Ddreena, and my seven-year-old Skye terrier, Ted, into the back of my old Toyota station wagon, its back end hunkering down under the accumulated weight. Ddreena's hind legs went numb the first day from lying too long in the same position, creating a horribly embarrassing and stressful situation when she rolled out of the back of the car, unable to gain control of her legs at our first stop for fuel. And even though I had tried my best to tie the plastic containers onto the roof of the car as tightly as possible, two lids blew away, the uppermost contents of those containers lost as they were scattered down the highways. On our second day on the road, a Sunday, the car declared itself void of oil as we passed through Rolling River. Most garages were closed, and the few that were open lacked the necessary part to repair the leak, but determination prevailed, and with a few extra quarts of oil riding shotgun beside me, we finally limped into the bunkhouse parking lot at Tarsa Lake.

I spent the following days completing the preliminary paper-work, meeting the tower supervisor, Bob, and the radio-room staff and trying to map out who was who in the main office. From there Valerie, another new recruit, and I were driven out to one of the

towers that was already open so we could ask Anne, the senior towerperson stationed there, any final questions.

The offices and bunkhouses in the Jordan district are set in a rural
area just off the highway north of town. Interpretive trails work
their way around the rustic wood-sided main office, tucked behind
about a dozen homes used by the forestry employees. The road continues past the houses and on down to Tarsa Lake, which was just
beginning to thaw under the sunshine and warmth of mid-May.
There was still plenty of snow in the woods. I wondered how much
remained at my tower.

There were four of us towerfolk staying in the bunkhouse at the
time: Valerie, who was on her way to Stoney Tower; Rick, who
had been stationed at Connaught last year but would be manning
Signant Tower this year, and Laura, who was returning to Vickerage
Tower. Although their exteriors often differ, the bunkhouses I have
seen have similar layouts that combine 8–10 rooms, each equipped
with two single beds, one large functional kitchen with several refrigerators to store the large amounts of food that towerpeople and
other staff bring with them, and a large living room with plenty of
seating. Forestry staff and fire crews also make short-term use of the
space over the season.

Valerie and I had roomed together during our training in
Hinton just a few short weeks prior, so it was wonderful to share
all the excitement of our impending departures as well as review
what we had learned in our course. I had actually met Laura back in
Calgary through a friend of a friend and had visited her home where
I put her to the test, querying her about so many tower things I still
couldn't visualize. It was nice to see her again.

We would all be going into our sites before the week's end;
Laura first, followed by Valerie and I, and finally Rick. So no sooner
had we settled into a routine at Tarsa Lake than it was time to pack
up and move out to our towers. On the big day I rose early and got
my gear and the dogs ready as Rick and Matt, the driver, helped

me pack everything into the big one-ton truck. Poor Ddreena was crated and loaded, and even though we placed her directly behind the cab with as much gear as possible piled all around her, each time I looked out the back window during our drive, she looked like she was caught in a hurricane. Add to that the dust and the potholes as we left the paved highway and made our way past Zephyr Lake, and she, without a doubt, had the longest day of any of us.

Nine years is quite elderly for a wolfhound. I wondered what impact my new career might have on her. I had imagined a quiet, peaceful summer far from the hustle and bustle of city life. But my imagination also summoned up images of altercations with porcupines or bears. It was one thing to offer myself to this new adventure, but I worried that I was placing my companions at risk. I wondered how they felt about that. Were they willing sidekicks, or would they have chosen alternative arrangements?

The job I had left to accept this newfound career had seen me supervising field technicians as they visited drilling rigs throughout Alberta, and on many occasions I had given out long, complicated directions into the Zephyr area. It was fascinating to see first hand what I had only tried to describe in the past. The first part of the road was well kept, the landscape flat as we passed large areas of marshland with an abundance of migrating waterfowl. Rough, overgrown, wet in some places and dusty in others, roads snaked everywhere. The landscape was a mishmash of pumpjacks, dead equipment, small gas facilities and gnarled black spruce. And all of this human detritus surrounded a lake that was a very important migration stop for myriad waterfowl. In just our one trip down the road, the diversity of the area revealed itself as a lynx crossed the road in front of us, a variety of coots and ducks sailed across shrubby, deadwood strewn ponds and sandhill cranes flew high overhead.

We arrived at the airstrip, which turned out to be past the townsite of Zephyr. The airstrip itself was nothing more than a bare, dirty hole in the woods created by bulldozers that had merely pushed back the forest, and onto which someone had added a number of aging closed-down shacks and leaking diesel fuel drums. Matt and I released Ddreena from her crate and helped her off the truck, and took a few minutes to look around. Ted got his feet wet soon after exiting the truck, and within minutes increased his total weight by half again from the clods of mud cemented to his long hair. I groaned as I thought that somehow the helicopter pilot wasn't going to find the mud endearing.

Just as we were beginning to wonder if we had indeed found the right spot, we heard a distant buzzing sound. The sound steadily became louder, and soon the helicopter was visible above the trees. As it descended, my companions ran for cover from the blowing dust—Ddreena and Ted straight to the muddy area, Matt to the truck—and I ran off to chase light bits of gear that were blowing away. Had it been true sand, the dogs and I could have created a beach at Connaught with what we carried away in our scalps and gear that day.

Since we had the larger of the two trucks, some of Valerie's gear had come along with us, so the first load went to Stoney Tower, which was only seven minutes' airtime away. On the return trip the 'copter brought back an empty 360-litre propane pig slung by a strap to its belly. It was very cool to watch the pilot place it ever so gently into the back of the truck.

Then it was finally our turn. Bob, our supervisor, was already aboard. The helicopter pilot was sweet, polite and handsome, and best of all, a lover of dogs. Luckily Ddreena's crate was too big for the helicopter, so we didn't have to use it. All 57 kilos of Ddreena was captured and heaved against her will into the aircraft, and we were off to the tower with all the gear that would fit in the remaining space.

The view of the land from the air was exhilarating. A perfect opportunity to get an overview of the geography and ecosystems below—primarily spruce, but more pine than I had imagined. There was still much snow on the ground and an amazing tiered series of ponds created by beaver. Some dams looked to be nine metres wide, with a beaver lodge in each pond. We travelled north up Sandal Hill, the Kimber Plateau visible to our west, more hills far to the east. What a feeling it was to crest a ridge and see the tower start to materialize in the distance, the cabin becoming visible through the drape of trees that surrounded it. The pilot was accommodating, circling Connaught as we approached so I could take some pictures before landing. It seems like there are hours of memories in that flight, and yet we were only in the air about 10 minutes.

We touched down and quickly unloaded; then the pilot was off to pick up the next batch of gear, leaving the dogs and me to open the cabin and investigate on our own. Standing in water ankle deep in the mossy yard, we took our first look around. The yard was larger than I expected. With the snow still melting and the frost only beginning to come out of the ground, the meltwater remained on the surface. The winter road, which Rick had told me would probably never be dry enough to walk far along, entered the yard from the south. A patio-stone walkway joined the cabin, the generator shed and the tower; and a wooden boardwalk led to the outhouse. The biff itself was pristine and situated in the prettiest spot in the yard, under a stand of mature pines that sang constantly with the breezes that day. That most peaceful sound made me feel immediately welcome.

The cabin at Connaught had remained unchanged since it arrived in the winter of 1973–74, but it had been treated with a gentle hand. The cabin and the generator shed were what in Ontario we call trailers but what Albertans refer to as shacks. The cabin sat on the sledge that was used to drag it in, allowing a good airspace between it and the wet ground. The cabin was definitely listing from the effects of freezing and thawing muskeg.

Steve, the fellow who had served 17 seasons here, was definitely "a place for everything, and everything in its place" type of

My first view of Connaught as we prepare to land on opening day.

person. You could do surgery in the generator shed, it was so tidy, all the tools mounted on a huge piece of white-painted plywood, the outline of each tool drawn with magic marker so there was no mistaking where it belonged. The previous towermen had made opening the kitchen cupboards as much fun as Christmas morning, with a motley assortment of emergency foods to investigate.

Soon the helicopter returned; the next job was to get all the propane appliances going (Bob couldn't get my fridge started, so the pilot ended up taking it apart and starting it manually, which took a bit of extra time) and test the radios.

There was one set of radios in the cabin; another interconnected set was in the cupola atop the tower. The setup of radios can vary from tower to tower, so there is no definitive diagram in the manual. At Connaught there were two separate sets of radios. One was for the simplex channels (used mainly to make contact with aircraft, requiring "line of sight" to reach the intended recipient), and the second was for the repeater channels, requiring you to "key up" (depress and release the mike button) before starting a transmission, because the signal first hits and activates a repeater that sends the signal on to the recipient or the next repeater. This allows for transmission over longer distances that lack line of sight. Bob first explained the set-up in the cabin, then we walked out to the tower.

"Up you go," he said.

I looked up. During training week, when the time came to climb a tower, it had snowed and the tower had iced up, so we never had a practice run.

I have always been afraid of heights (despite my penchant for climbing trees). A poor sense of balance left me hesitant on exercises such as this.

I looked at Bob, and then I looked up at the cupola again. Every bit of me agreed there was no going back now—and up I went.

Just like that. It's funny how dreams-come-true can do that. I was winded and needed a few moments to catch my breath, but the climb went off without a hitch.

It was exhilarating! As I crested the tops of the trees, the images I had seen from the helicopter became visible again, and the full expanse of my new world settled into my brain. Once into the cupola, I tried to describe to Bob what radios I found there. After a few attempts, it was clear that we weren't able to give accurate descriptions to each other. I suggested he come up so we could discuss this more effectively. He declined the invitation, explaining that he was afraid of heights.

The final few details discussed, the last of my gear having arrived, my supervisor and the pilot were on their way. As I watched the helicopter rise from my yard and disappear from view, I looked back at the pile of gear sitting in the yard. Ten five-gallon containers and six flats of bottles that composed my stock of potable water still needed to be relocated to the shed. All my personal gear needed to be moved into the cabin, then the rain barrel had to be filled with snow—it was a very long and tiring day by the time all was said and done. As for Ted and Ddreena, they were sound asleep in the yard within a half hour. I took a moment to enjoy their peaceful expressions and thank the fates that we'd all arrived together, safe and sound. My regrets for putting Ddreena through such a difficult day began to fade; a smile began to grow as I thought of what a wonderful summer they both had ahead of them.

It seemed so very long before I had things organized enough to think of the site as functional, but I finally felt I was ready to call in my official opening message to the office. I swallowed the lump of fear in my throat, keyed up the radio and raised the fire centre, announcing formally that Connaught Tower was now officially open for the 2002 fire season. Accumulated hours on the generator and propane tank fuel levels were duly reported as well.

I managed a bite to eat and found the dogs their dinner, but was too tired to even chew. I did my best to make the single-bed-sized bedding I had brought with me fit the double bed. There was still light in the sky when I fell into bed around 23:30.

"This is XMR 426, open for the season."

Home Sweet Home

The alarm sounded at 06:30. I only needed to move a few times to feel the effects of the previous day's labours. It was going to take a big effort to limber up. I headed out for my first walk of the day to the outhouse—it was fun to have the freedom to walk around outdoors in my pajamas and duck boots with nary a concern for anyone being around.

During one of our talks back at the bunkhouse, Rick had told me that the largest creatures he had seen at Connaught last season were rabbits, and sure enough, when I opened the door that very first morning, there was a grand fellow just at the foot of the stairs. Dressed in a patchwork of summer and winter attire, he seemed a bit surprised to find the cabin occupied again and even more surprised to see the dogs. A short chase ensued, but the hare was never in danger, because the dogs gave up at the edge where the site met the forest, about nine metres away from the cabin door.

As I set about making coffee and thinking about how I could organize the day, I took another look around my new home. The kitchen table and chairs were of the older chrome-legged variety and located under the east window. With the bedroom at the north end of the cabin and the office at the south end, this middle portion I was standing in could most aptly be described as the living area. The kitchen cupboards, sink, fridge and stove occupied the west wall; the table and propane heater stood against the east wall. The fridge and stove were smaller than the city variety, and I made a mental note to plan how to best utilize the refrigerator space. It was also difficult to tell how quickly the propane fridge cooled, since there was no whir of a motor. I had to remember to check at regular intervals to make sure I had it set at a workable temperature before I lost any of my month's supply of groceries.

The trailer sported two doors—one from the centre of the east side, the other exiting the west side from the office, where it was only a few steps to the ladder at the base of the tower. The windows were of the sliding type, and all the screens were intact, thank goodness. The walls of the cabin were depressing, because the entire cabin was covered in a dark-coloured pressboard panelling, making everything dreary. In an attempt to brighten it up and add a bit of interest, Steve had left a good assortment of Kananaskis Country and bird identification posters.

It's funny the habits we get into. For those first few days I would repeatedly reach over the sink to turn on the faucets ... which weren't there. A bucket of water drawn from the rain barrel and set by the kitchen door was the source of water for any need other than drinking. So now I would need to walk over to the bucket and ladle what I needed. The drain that ran from the sink deposited the grey water on the ground just outside of the cabin. Although this was a serviceable arrangement given the amount of water I would be draining from the sink, I took seriously my responsibility to defer to environmentally friendly soaps and keep tempting aromatic residues such as bacon fat to a minimum.

I also needed to revamp my bathroom routine. City washrooms allow for toileting, bathing and miscellaneous washing and primping. Here the toilet was outside at the end of the boardwalk, and the bathing area was yet to be designated. Face and hand washing was done at the kitchen sink, which still left primping. Hmmm—where does one place toothbrush, lotions and other toiletries that are normally housed in the bathroom cabinet? The top of the chest of drawers in the bedroom was thus rearranged.

The bedroom was appropriately placed at the north end of the trailer, allowing for as much darkness as possible. Rick had hung dark-coloured garbage bags over the doorway into the room and over the north-facing window in an attempt to keep the long northern days at bay. The double bed was complete with headboard and a side table on which I could sit my lantern safely. A good-sized closet, a low-boy style chest of drawers and an extra pantry cupboard filled the east wall. I had brought along quantities of supplies that would hopefully keep me the entire season, so it didn't take me long to make full use of that space. The bed was placed against the west wall. There were no chairs except for the kitchen and desk varieties, and since the trailer listed to the west, I would lean back against the wall, my feet extended across the bed, creating the closest substitute I could to a reclining armchair.

The office was a pleasant space containing a big desk filled with enough of this and that to pique my curiosity and inspire me to take inventory. I looked around at all the things Steve had left behind in this cabin: a raincoat, boots, pictures, the posters on the walls, the notes left for his successors, the visitors' log from 1978 to present. I also found little treasures such as a notebook full of notes from a course he'd been taking in Gaelic, with a quote printed neatly and pinned to the windowsill. There were also years of neatly handwritten weather records. I almost expected the door to open at any time and Steve to walk in, this space seemed to be still so inhabited by him. Desk drawers and a well-constructed storage cupboard held enough little bits of string, screws, miscellaneous bits and batteries to make me feel comfortable that I could tape, glue, repair, oil or tie up any old thing that might come along.

I also took time that day to set up my computer and make my first attempts at dialing the modem through the cellphone that was installed both on the ground and in the cupola. This particular style of phone, what is referred to as a "bag phone," is actually just a muscled-up counterpart to the cellphones most people carry in their pockets (three-watt vs. one-watt). There was a cellphone tower just at the south end of Sandal Hill that allowed for an excellent connection. After a couple of minor changes to the modem settings, I was able to send a short note off to my mother and some close friends, letting them know that all was well.

As I hung up the phone, I looked around, surprised at how much I wasn't getting the feeling that I was roughing it. Roughing it is kayaking on Georgian Bay, pitching your tent every night, digging your own latrine pit each time you feel the need, working out a food schedule that requires no refrigeration. Here I had a phone, fridge, stove, heat and a washroom, even if it was a bit of a walk from the back door. My groceries would even be delivered! Is it really roughing it if you have the Internet?

Later in the morning my supervisor called. He began apologizing, informing me that when the helicopter had returned to Jordan and they had cleared the last of the gear out they had found one of my boxes. What with all the confusion of opening day, I still hadn't discovered one was missing. I asked him if he could look inside and tell me what was in the box. He replied with a list of books and yarns and such.

I sighed as I thought about the contents of that container. It was the bin in which I'd packed all my leisure activities, including my various identification guides, my reading, all my knitting and embroideries. Not much I could do at that point; I surely couldn't expect a helicopter, at $800 to $1,000 per hour, to make a special delivery. We came to the agreement that my container could come out during another mission, whenever that might be. Luckily, a few newspapers had been packed into a different box, and during my last trip to the hardware store in Jordan I had purchased a jigsaw puzzle on an impulse.

I admit to pacing at least a couple of circles around the kitchen at that point. That jigsaw puzzle was going to be in for one heck of a workout.

By the end of that first day, I had climbed the tower five times, and spent most of the morning collecting snow to fill the water tank, topping it up again late in the afternoon. The dogs and I had tried to take a bit of a walk, but it was too wet to wander.

Each time I climbed into the cupola, I found the view no less amazing. I could discern the Kimber Plateau in B.C. to the west, flat muskeg plains to the north, then Dene Lake to the east, with the McLellan Hills rising behind it. With a last leap of imagination, I could imagine Corris Tower and the highway to Yellowknife lying beyond them. I could easily define the path of the Connaught River, which meandered a mere six kilometres from my doorstep, carrying the waters of Dene Lake to British Columbia; the bare aspens lining the banks showed as grey veins, contrasting with the predominantly spruce-green landscape. Sandal Hill hid details of the near landscape to the east; the top of a similar hill (except for a cellphone tower) was all that was visible until my view came around to the southwest and the valley between Sandal Hill and the Kimber Plateau appeared once again.

With the tower site very close to the northwest edge of the hill, the 183-metre boost in height enhanced my view of the boreal muskeg. Many little ponds and small rises gave definition to the muskeg below me. All these landmarks were useful to me, since most waterways were marked on the map in the cupola, allowing me to use them to gauge distance. Since I have never had an accurate eye for distance, I was grateful for that.

I have never thought myself to be very imaginative, but, oh, the daydreams that filled my mind as I looked out over these expanses! To sit atop the world where there wasn't likely a human as far as the eye could see, to look into two provinces and a territory in a sweep of one's eyes, imagining what the landscape was like beyond my view. How far could a bear wander in a day down there, how many deer shared these forests, how many of each were within my primary area? Did they cross paths, and if so, how often?

In the days ahead I would watch small flocks of sandhill cranes cruise by my tower, intent on their mission to spend the summer in Alaska. I could watch them unassisted as they passed the tower, then with my binoculars until they finally disappeared in the distance. Imagine them moving about as they were meant to do, no hunters arguing that the flocks needed thinning, no homeowners complaining of damage to fields from surging populations. With a minimum of human encroachment jeopardizing population dynamics, the creatures and natural systems below me were breathing, living, flourishing and passing away naturally.

Did this mean that my view contained no signs of human disturbance? If only that were so! I would often close my eyes and imagine that below me there were no cutlines, that the landscape was not criss-crossed with the paths left by seismic testing. However, such is Alberta's passionate search for oil and gas that no corner of the province is left unscathed. Between my perch on Sandal Hill and the Kimber Plateau I could see one little gas plant, and I knew there was another plant blocked from my sight over on the slopes of the McLellan Hills to my northwest. On my map, a lodge and a Native reserve were marked along Dene Lake, although I learned later that there were no longer any permanent residents living on it.

So there it was laid out before me: my work world. No matter how often I closed my eyes and opened them, all that beauty of nature was still there; I wasn't dreaming. I often pinched myself those first few days.

I was placed on low hazard level, for which I was grateful. I wasn't even able to report "snow gone" yet. I continued to fill the water barrel with snow each time I noticed that the melting process had created more space. Once the remaining snow dropped to where it had become "patchy to the extent that the observer [could] walk around the snow patches for about one kilometre from the observation site," then my opening indices could be calculated and set by the fire centre. Time spent in the cupola would be dictated by the hazard levels calculated from my weather observations.

The dogs and I entertained ourselves by investigating the site. The cleared part of the yard—the lawn—was a mix of grass, moss, willow, aspen shoots and a variety of wildflowers. The wild strawberries had already started to green up as the snow melted and they became exposed to the sun. Beyond the yard the deciduous and coniferous trees were still at closed bud stage. The Labrador tea was looking typically rhododendronish for this point in the season, with the overwintered leaves dull, curled and drooping. The mosses were in their glory what with all the meltwater and warm sunshine. Rick had worn a jogging track around

Cupola Occupancy Guidelines by Hazard Level

All Levels: At morning weather and at sunset.

Low: Observations every two hours from 10:00–18:00 hrs; no continuous cupola occupancy.

Low–Moderate: Every two hours from 10:00–12:00 hrs; cupola occupied 14:00–19:00 hrs.

Moderate: Hourly from 09:00–13:00; cupola occupied 13:00–17:00 hrs.

High: Hourly from 09:00–11:00; cupola occupied 11:00–20:00 hrs.

Extreme: Cupola occupied 09:00–20:00 hrs.

the edge of the site during his stay, which the dogs and I quickly adopted as our path for walks as it was solid, ice free and above water level.

The more we walked the path, the more I began to see the organization of the south end of the yard. We located the little vegetable gardens, wrapped securely in chicken wire to keep the bunnies out, and two large covered pits that Steve must have dug. I never did figure out what Steve used them for, other than perhaps burying burnt garbage. But each pit had a floor of clean earth, and I really had no urge to dig around and see what I might find. Sometimes things are best left unknown.

I had noticed that this area had been thinned and organized, but it took a few walks before the enormity of it finally sank into my brain. In his many years at Connaught, Steve had crafted an arboretum! Just Steve and his hand tools, over all the years he was here. I have often tried to estimate the actual size of the area, perhaps around 180 square metres, but every inch held beauty. Using only native species found on the site, Steve had created something truly magical. As different species greened up, more paths became apparent, meandering among the perfectly shaped trees, with clumps of Labrador tea and wild roses, all carpeted in bunchberry. Mosses and lichens in a full range of colours nestled in shady places, and low shrubby plants from the wild blueberry and cranberry families covered the ground with their shiny green leaves and red stems. Just within those first few days, many of the mosses and lichens had taken advantage of the weather and "bloomed"—each in its own peculiar way. The equisetum practically leapt from the ground; the Labrador tea unfurled itself, raising its leaves to the sun. Each time we walked through, the dogs and I would take a different path and find another entrancing aspect; the magical combination of light, composition and colour could rival a Monet painting. The area was divided into an east and a west section by the winter road that entered the south end of the yard. I often found myself in those early days standing on that road, arms wrapped around me, dogs at my side, turning slowly and drinking it all in, and considering myself fortunate for all this beauty.

So passed my first few days. I would be up early, pass morning weather at 08:00, take a quick run up the tower before I walked the dogs and addressed some sort of chore, climb back up to the cupola starting at 10:00 and then every two hours until 18:00, reporting afternoon weather at 13:00. The fire centre did a final check of the towers at 19:00, and then the rest of the evening was my own.

I often ended up eating "high latitude" late dinners because the days this far north were already longer than I had ever experienced. Even though I was sure that it wasn't time to do our major vegetable planting yet, I did get the herbs planted in the raised bed along the side of the cabin, since they were easy to cover at night. The African daisies and other bedding plants were shuffled in and out of the cabin as the weather allowed.

Yes, Connaught was beginning to feel like home.

Off to Work I Go!

The first item on the daily tower agenda is weather, so on that first Connaught morning I was up and ready to tackle the assignment. I allowed myself extra time at the Stevenson screen to think my way through everything carefully, because it does take a bit of time to convert the classroom teachings and notes to the actual process. The Stevenson screen consists of a louvered white-painted box that holds a series of thermometers, one which gives the maximum temperature of the previous day; another that gives the overnight minimum; and the "wet" and "dry" thermometers that make up the psychrometer, which allows the calculation of dew point and humidity. The design of the box allows for the most accurate temperatures without the effect of direct sun, wind or rain, so that one can accurately compare data from many tower sites. By the time the fire centre calls, maximum, minimum and present temperatures, dew point and humidity, wind speed and direction, sky conditions and cloud types have to be recorded and ready to pass along.

I was just walking in the door with my data when the radio keyed up. It was Valerie from Stoney asking me for a cross shot on a smoke she had spotted (fire-tower observers are responsible for detecting and reporting smokes—at which time a forestry officer will investigate and change the smoke report to a fire report

Weather Information Collected by Fire-Tower Observers
(Approximately 08:00 and 13:00 daily)

<u>Max and Min Temps:</u> Reported at morning weather only; pertains to previous day, in °C.

<u>Precipitation:</u> Rain (mm) measured in rain gauge and/or snow (cm) depth measured with a ruler.

<u>Wind:</u> Anemometer mounted somewhere on site gives wind speed. Direction and gusting also reported.

<u>Relative Humidity:</u> Read from the fan psychrometer housed in the Stevenson screen. This also calculates dew point.

<u>Sky conditions:</u> Calculated by mentally dividing the sky into ten equal pieces and deciding how many tenths contain clouds. Recorded as clear (0), scattered (1-5), broken (6-9), overcast (10).

<u>Visibility:</u> Calculated by asking oneself the question, "At what distance would you be confident of detecting a smoke?" The primary area for every tower/lookout is 40 km, but visibility can be limited by smoke or fog, e.g. Given in kms, perfect visibility would be recorded as "40."

<u>Clouds:</u> Using the tenths recorded when giving sky conditions, how many of each contain clouds that are either high, middle or low? Special attention is given to describing cumulus (storm) clouds.

<u>Present weather/obstructions to vision:</u> What the weather is doing at that moment—rain, thunder, lightning, fog, haze smoke, etc.

if necessary). So much for being ready for weather! My data, coffee and cereal were left on the desk, and I headed up the tower, determined to assist her in fulfilling the fire-tower observer's mandate to report all smokes within five minutes of discovery and while still under 0.1 hectare in size. I got Valerie's bearing from her just as the radio keyed up and the towers began relaying their weather to the fire centre. When it came time for what would have been my maiden report, I could only apologize and tell the radio operator that I would radio the information as soon as I was able. Using Valerie's bearing and distance to estimate the location of the smoke, I nervously turned my binoculars to see what I could find. The area she had indicated turned out to be blind to me, because it was off the southern edge of Sandal Hill; so if I were to spot it, I would be looking for something drifting up from the edge. A few moments later Valerie radioed over and decided that the smoke was actually a gas flarestack—problem solved. Once the adrenaline rush was over, I took one last observation then came down the ladder, radioed in my weather and sat down to my breakfast.

The settling-in process continued. What with getting into the new routine, becoming more comfortable with climbing the ladder, contemplating how, where and out of what materials to build a shower stall, coming to grips with my radio shyness, growing more confident with my weather observations and trying to learn the lay of the land to be ready for whenever that first smoke came up, there was no time to be bored.

The weather gradually shifted from sunny and warm to overcast, bringing us some sorely needed rain one morning in the third week of May. Tom down on Zephyr Tower was calling in his 47th or 48th smoke report of the season, and Connaught hadn't been open a week yet! During weather reports, the fire centre dispatchers updated us on the fire situation in other parts of the province—there was already a 13,000-hectare fire out there somewhere, but I kept missing the location.

Zephyr's abundance of smokes can be attributed to the fact that it is a "settlement" tower; it overlooks an area settled by humans. With this comes a much larger workload than at a "lightning" tower such as Connaught, where most fires are started by Mother Nature.

At towers such as Zephyr, the towerperson has to keep track of "permanent smokes" such as industrial stacks or garbage dumps as well as keep a running list of current fire permits.

In Tom's case, he was looking over an area containing hundreds of flarestacks and inhabited by an Aboriginal community. Although he had many seasons as a towerperson under his belt, it was his first season at Zephyr, and he undoubtedly found each day a challenge as he tried to distinguish all those flarestacks from honest smokes. Although towerpeople are never disciplined for incorrectly calling a smoke (you could say that brainstorming rules apply—no one is ever wrong), it's a matter of pride for the observer to be able to tell what he/she is looking at. Especially when one considers the expense of helicopter usage, it's just as much of an achievement to successfully not call a smoke that isn't a concern to forestry.

It was interesting to watch the weather conditions change as the storm got closer that morning. The relative humidity rose, the wind changed direction to south-southeast and the cumulus clouds began to build and tower.

Weather reporting had been another worry of mine before reaching the tower, but it was surprising how quickly I learned to read the skies. On this particular day the clouds only brought a steady drizzle. It was easy to tell whenever the rain stopped because almost immediately the yellow-rumped warblers started to sing. It was amazing how quickly the water level could rise in my yard. Muskeg indeed!

I made the climb up the tower once that morning, but stopped halfway up when there was just nothing to see—greyness stretched out in every direction. So midmorning, I decided to give Steve, who had spent so many years at Connaught, a call. Rick had suggested this—he had done the same at the beginning of his season. I imagined an aged, gravelly but robust voice would answer the phone. Steve was an ex-army fellow, a legend in the tower world. Yet his voice was frail and much more elderly than I ever would have guessed. I explained who I was and had a lovely conversation with him. I assured him that all was well here at Connaught, that his presence was still here in so many ways, that Rick had left everything in fine form. Steve asked me if I would be participating in a wildflower census for which he used to collect data. I promised him I would look into it.

The rest of the day was filled with more chores and further investigations in the arboretum, camera in hand. Along the jogging path on the east side, I noticed several piles of scat. They had been left by an obviously good-sized ungulate, but resembled that of neither deer nor moose. I thought they looked fairly fresh, but the lack of footprints around them made me think they had settled as the snow melted. I believed that it was too far north for elk. An email sent to a biologist friend brought me back the reply that it be might be woodland caribou.

I had just finished my supper when the phone rang. A very excited Valerie was on the other end of the line telling me that a bear had just come knocking on her door. A large cinnamon fellow had wondered who was playing the music and cooking such a yummy-smelling dinner. Upon turning about and seeing him peering at her through the screen door, she turned up the music and ran for her frying pan, but the bruin had made the wise decision to move along on its own (I would have run from the Backstreet Boys as well!). I found it funny when she told me that when she first heard the noise at the door and looked up, she thought it was Ddreena come for a visit. When Valerie saw the paws on the door, she realized that if it was Ddreena, she was in sore need of a pedicure.

There was one last trip up the tower at sunset that day, but at the 26-metre mark on the ladder I stopped. Any higher was a waste, because the clouds were still far too low to be able to see anything. But evening vespers were being sung by the warblers as the setting sun fell through a hole in the clouds, turning the sky brilliant pink and the hazy landscape below a vivid lavender. Just below me, the anemometer was dead still. I just hugged the ladder and listened and watched and absorbed the peace.

Midnight approached, and it was time for bed. Ddreena was snoring away on the kitchen floor, Ted was in his little den under the desk and there was still light in the sky.

The following morning greeted us with enough frost to make the muskeg crunchy and hug my boots with every step. When I opened the front door, I clucked at myself for forgetting to bring in the plants the night before. A closer look revealed that my beautiful, sunny gazaria was no longer—now just a stump in the dirt—and half my leeks were pruned to ground level. Darn those ungracious bunnies, to come right up on my porch and snatch them like that! I made a mental note not to repeat that mistake.

I was disappointed to lose the gazaria, but if the plant was that tasty, then it was doomed over the summer in any case. As for the leeks, well, they'd be back up again in a few days. Luckily, the more expensive African daisies seemed to be less tasty, but not wishing to push my luck, I decided that a fence should go up around them. Then Ddreena walked out on the porch, bumped into my pot of corydalis and dumped it. That particular pot had accompanied me ever since my senior research project at the University of Guelph, where I picked the original plants out on the Mink Islands of Georgian Bay, and I had been determined to have them here with me and see how they liked life on the muskeg. I thought it a miracle that the pot had arrived intact, and the seeds had just begun to germinate. I carefully sorted through the dirt and found three of the tiny seedlings and did my best to rescue the soil and replant them.

General projects continued. I got out the homemade ladder and sealed the eavestroughs in a few places, beat down a few nails in the boardwalk, which had been heaved by the winter, and started building my shower. The wind turned to the north-northwest and blew all day, a colder wind than the past few days, but I was finally able to call "snow gone" in my weather report.

It was good to see the world from my tower again as the grey haze from the previous day blew away with the wind. A flock of yellow songbirds floated by on the breezes; a hawk soared on the updrafts blowing up the hill's side just to the west of the tower until it was only a speck in the sky. In all the time I watched it, I don't think it flapped its wings once. I could only imagine his spectacular view as he sailed over the country to the west of the hill, probably close to 300 metres above the valley.

I have come to appreciate that at towers, sponge baths suffice, but oh, to stand under a hot shower! My engineering ideas slowly came together, and the shower project finally got underway. I had great fun sorting through the stacked spruce saplings Steve had left and rummaging through the sheds to scavenge nails and bits of wood. As for location, I decided that I would take advantage of the tall lovely stand of pines that sheltered the outhouse.

I began with a pallet I had found behind the fuel shed. I was grateful for the sharp axe as I rabbitted the spruce poles and nailed them to the four corners and midway on the long sides, a few smaller poles used to brace and square them. I had found two small pieces of plywood, and managed to fit them between the corner poles to form a backboard. Although the pallet was missing a few "floor boards," I had found some sticks over by the vegetable garden that would work perfectly, because the bark had been peeled from them and they had weathered to a smooth finish. Steve must have used them to stake up plants. I rabbitted each end of the sticks as well and nailed each to the pallet, allowing a serviceable floor that would allow me to stand a few inches off the ground but let the water drain through. The finishing touch was provided by a sapling that had been bent over by the snow and had remained U-shaped. It fit perfectly, forming a halo just above my head to which I tied the tarp that would enclose the three open sides. I added a small scrap of two-by-four to hold assorted bottles, a nail on which to hang my facecloth, and it was done. Between weather and trips up the tower, it only took me close to a week to complete!

During those same days, Valerie received birthday greetings from our community of towers over the radio. She thanked us and invited us all over for cake. Funny girl!

My first attempt at building a shower stall: an engineering marvel for a beginner like me.

Neighbours Stop by for Tea

Even at the end of May, the nights at Connaught, just 50 kilometres south of the Northwest Territories border, hovered around -5°C, and the days were anywhere from a paltry few degrees to the mid-teens. What I had often read about the north was true: each day huge strides are taken by the inhabitants of this latitude. From the growth of plants to the changing colours of hares, the seasons march at a quicker pace in the north.

One morning I was enjoying my breakfast and gazing out the window, when who should saunter into the yard but a woodland caribou. I ran for my camera, but, afraid to go outside and startle him, I stayed in the cabin. About midway across the yard, he started to act skittish. I thought that he must have seen me moving through the window, or that my camera lens was glinting in the sun, but it wasn't until he loped off through the north end of the yard that I remembered Ddreena was outside. I found a wolfhound with the most incredulous look on her face (I'm sure her mouth would have hung open if that were a possible canine response).

Later in the morning the dogs and I went out to investigate the caribou's path; I wanted to see its hoofprints. We walked up the cutline a bit, then followed the path he took through the yard, to no avail; not a mark anywhere. Such is life living on spongy moss, I

A woodland caribou surprises us with a visit.

supposed. I had found some information that indicated caribou eat Labrador tea as an early season delicacy, so now the scat that I had found previously and this appearance in the arboretum all made sense. I hoped it would come back for another visit.

Three days had passed since I had declared "snow gone," so the following morning I received my indices from the office. My hazard level was immediately upgraded to high, so I went into my cupola for the day. I was up the ladder just past the mark of 09:00. Not bad, considering that, although I had woken up in plenty of time for the weather report, it had been one of those "wandering" mornings (any 40-something woman will understand this: start one thing, wander onto the next, then onto something else). Before I knew it, the dispatcher was on the radio asking Tom to start us off with the weather reports.

"Nooo!" I shrieked, running out the door in my slippers, the dogs thinking I had lost my mind as I tried to look up and take notes on the clouds and wind direction while starting the fan on the psychrometer. (Connaught was the 5th in line of the 13 towers to deliver its report, so there wasn't much time.) Info in hand, I sprinted to the door only to miss the step and take a tumble on the steel steps. The final scramble in and the data transferred onto my report sheet, I only had to beg forgiveness for not having my dew point and relative humidity ready.

Time ticked slowly by in the cupola with nothing to keep my hands occupied. Not only did I miss my knitting and embroidery, which had been intended to fill the time in the cupola, but I hesitated to bring the little bits of reading I had upstairs because reading tended to make me drowsy. It wasn't in the script for me to start my career as an observer by sleeping through my first fire. The little wind-up radio that I had brought along to chatter to me during the days was a challenge as well. I was far enough away from civilization that the only radio stations I could bring in were all of the international shortwave type. A day of radio began with a few minutes of Radio Austria or Radio Netherlands until they signed off, followed by a Spanish-speaking Central or South American station until around 12:00. CBC radio appeared and then disappeared just after lunch, then an hour or so of a hilarious Washington DC station that espoused so much propaganda that some days I all but fell off my stool from laughing so hard. Once that station began to fade, I caught the news of the future from Radio Australia or New Zealand (they were giving the news from the next day, but they never seemed to know more than the rest of the world). Although I couldn't understand the words, the South American music was by far the cheeriest and filled the mornings well.

Even though the overnight temperatures still dropped below freezing, I decided to test my luck and plant part of my garden. I double-checked my seed packets, and their notations agreed that spinach and snow peas could be planted as soon as the ground could be worked. Part of my inspiration was no doubt the green thumb itch nagging at me; the other part was the fact that it looked like it might rain that night. The northwest wind had brought many clouds from the Yukon, and although there had been much virga (visible rain that doesn't reach the ground as precipitation) floating past, only a very few drops spattered the cupola windows.

The dogs actually looked happier those days that I was in the cupola longer. Perhaps their familiar city routines were soothed when I left for work each day; or perhaps they were simply craving any familiar routine at this point.

I could now differentiate among the three bunnies. The first was almost all brown now except for the tips of its ears; the next still had white on its belly and more on its ear tips than the first; the third had a big white spot on its nose to go with its white underparts. I noticed that they had adapted to the dogs' presence since Ddreena could now share the yard at length with them.

Bird identification was an ongoing challenge, what with the lack of guidebooks. Thank goodness that Valerie had packed one, and that I could contact various birding friends via email. With assistance I had been able to determine that the two similar but distinct warblers I'd been seeing and hearing were both yellow-rumped, but some were of the Myrtle race and others of the Audubon race. A sapsucker flitted around the edge of the yard, and although he picked and poked under the bark like his woodpecker cousins, he didn't run up and down the tree trunks as much, but spent his time out on branches.

The late spring weather over the next week was exactly what I love about Alberta—constant flux. One day it would be warm and glorious, a couple of days later it would be snowing. A cold front would blow in from the N.W.T. and from a cloudless sky vast sheets of horsetail clouds would move in, puffy cumulus below them, casting galloping shadows across the muskeg plains. But no matter what the weather gods heaped on Sandal Hill, it didn't stop the world around me from marching onward to summer. Under huge clear blue skies, the buds on the aspens were becoming brave enough to extend their new leaves into the world.

As Valerie and I got to know our individual ecosystems better, we became aware of how different things can be, even over small distances. Stoney Tower was only around 80 kilometres south as the crow flies, but while Sandal Hill might be around 17°C, Valerie would report 24°C, which she found stifling. Connaught's hilltop location allowed for an almost continuous wind, which moderated the temperatures. While Stoney's aspens were in full leaf, mine were just contemplating peeking from their buds. Looking down at the aspen groves off the hill to the north, I saw those once-grey groves were now crowned in a faint aura of gold.

As I had more time to ponder the yard and the surrounding area, new questions arose. There seemed to be more caribou scat where I had first noticed it in the yard. I decided that must be the spot where he napped at night, but I hadn't had an actual sighting to confirm that. Another thing I contemplated was the number of "paths" around the yard (other than Rick's running track). Back at the Ontario farm where I used to board my horse, there were all kinds of paths that the coyotes used, since they are creatures of habit, just like us. When I first saw the caribou scat, I presumed that some of the paths that entered the yard were made by caribou. But now, as I discovered more and more of them, I decided that many just weren't big enough for a caribou to pass through. Then I found an aged pile of wolf scat in the same area of the yard as the caribou scat—just your traditional kind of wolf/coyote scat, primarily rabbit fur, but this bunch had a nice assortment of bone bits to make it more interesting.

I hadn't seen the bunnies during the past few days, and having caught the dogs sometimes staring into the woods with perked ears, I took it upon myself to scan the thin forest surrounding the site to make sure I didn't see anything lurking about that might enjoy a quick dog lunch.

I never did see anything the whole season. The forest and its inhabitants went about their own business, and we went about ours.

Of Harriers, Storms and Tests

As they have done for as long as the muskeg has graced the north, the aspens below the hill obeyed the annual cycle and began to leaf out, each clonal group acting in unison. A pair of robins charmed us with their morning and evening songs; juncos, ruffed grouse, Townsend's solitaire and other unidentified songbirds also became part of Sandal Hill, whether for a day or for the nesting season. This created the sweetest chorus of romantic song, whether they were singing the praises of the sun, the rain, or earnestly trying to entice a mate.

One particular day-after-a-rainy-day brought us a heady mix of western breezes, blue skies, cottonball clouds and the songs of the pines. Later that day I made the decision that it was finally warm enough: I installed the blue tarp on the new shower stall and had my very first hot shower since arriving. It still wasn't warm enough to really be comfortable wet and naked outside, but the space inside the stall was small enough to keep drafts out. It was glorious!

The pussy willows began to grow old and fuzzily golden, leaves started to extend from their buds. The new grass was now taller than the cured remnants of last year. It seemed that a new plant species arose from the yard daily. The bunnies had returned, but I couldn't tell them apart any longer, because they'd slipped finally into their slick brown coats of summer.

I had just settled in the cabin when I heard a rumble outside: thunder! I ran for my coat and gloves and climbed as quickly as possible up the tower. A storm was moving from north to south along the east side of Sandal Hill. As I watched it (and listened to Corris and Vickerage towers call in "first strike" reports as the storm moved through their viewing areas), another storm appeared from the northwest, moving directly toward me. Ah, life's challenges. My first storm watch and I would be watching for strikes moving in from opposite directions! The storm on the east side was difficult because the slope on that side was blind to me, making it difficult to tell where the lightning was striking. The storm from the northwest, however, was in full view, and I could watch it perfectly. As I was endeavouring to watch them both, an additional little cloud with an attitude decided to douse my tower in hail, increasing the difficulty as water and hail streamed down the windows.

The northwest storm front came directly toward the tower, but upon reaching the hill it came to a full stop less than a kilometre in front of me (was it contemplating sending a bolt at me, I wondered, in this true David and Goliath moment, as previous sins ran quickly through my mind). It then split before turning south, moving between my hill and the Kimber Plateau and sending the calved part around the north end of my hill in an attempt to catch the eastern storm. Between these two new curtains of grey and rain a new thin section emerged, thinning further to become a gauze-like golden veil, back-lit by the setting sun. The landscape had little definition, looking like a desert of golden sand, with only the shadows from small hills for texture. A muted blue sky with puffy cottonball clouds became visible in the distance, but still cloaked in that misty gold aura. I stayed at my post long after what was required that evening as the skies continued to taunt me, but with those beautiful misty images brushed over the broad landscape in sunset tones, it wasn't a hardship.

Moments like these began to accumulate. I consider one of the greatest benefits of being a towerperson to be having the time many only dream about to just sit and watch the glory that is Mother Nature. It doesn't happen at the snap of the fingers, unwinding from city life. It takes time and a true desire for that patience to enter your life. There were plenty of hours I spent fidgeting in the early days, wanting to pick up the phone and call neighbouring towers to while away the time, pacing around in the cabin with arms crossed, hands tucked in my armpits. But the habit of having to rush from one chore to the other was gradually replaced by the ability to slow down and appreciate both small images and large vistas, to inhale deeply through my eyes and appreciate the earth as she was meant to be.

I was busy "mowing" the lawn. Actually, I was busy snipping it, because the only tool I had to accomplish keeping the grass under control was a long-handled pair of hedge shears. There was a huge towering cumulus cloud directly to my north. It was decision time again. Should I stop what I was doing and run up the tower for a storm watch? I chose to get the compass from the cabin, so that if a first strike came I could at least take a rough bearing from the ground, and meanwhile continue with my chores. But it was a stubborn cloud, and I was beginning to believe that it would inevitably strike when my back was turned. So after about 10 minutes, I gave up and climbed the tower.

I was watching the storm's progress when I noticed a northern harrier working his way through the forest at the north end of the hill. Although a friend has shared her doubts that it could be a northern harrier, because they are more of a prairie denizen, I had no doubt because this fellow was the ghostly silver colour of a male. There is no other species of hawk with which it can be confused.

Unlike other hawk species, they do not climb above the forest appreciably; rather they stay closer to the ground, where they deke and dodge among the trees (when not taking advantage of the cutlines), which seems like a lot of work for a hawk of their size.

By comparison, the kestrel I had watched as he flew high above the forest seemed to soar so effortlessly. I was just thinking that I'd never observed the harrier rise above the treeline, when he suddenly turned upwards, sailing on the storm updrafts from the edge of the hill. He spiralled upward, never needing to flap his wings. At one point I took the binoculars away from my eyes only to realize he was so high that I couldn't see him anymore unaided; he was almost to the top of the storm cloud. Just sailing.

Although I managed to find him once or twice more with the binoculars, he finally sailed out of my view, off into the clouds hundreds of metres above the muskeg.

After running down to pass my weather report and grab a bite to eat, I was back up the tower again. Six storms awaited me. Two each to the north, east and west. Once again they decided that the hill was too much of an effort to climb; they'd rather go around, each one off in an independent direction. After pondering all six, I decided the western storms had the biggest potential for lightning, being the darkest and most forbidding. I settled in to try to wait it out. After staring them down fruitlessly, I decided to do a "full round" observation.

There it was to the north ... a smoke! An honest-to-gawd, straight-forward, white, billowing smoke. Ohmigosh! I had sat wondering these past few weeks if I would recognize one if I saw one, but there was absolutely no mistaking it.

I couldn't believe how quickly it had come up! The adrenaline ran, and of course all the obvious ensued: I couldn't find my detection report forms, then I couldn't find a pen. Should I call in a pre-smoke first? Should I try to get most of the info, take a bearing from the firefinder (an instrument that measures the azimuth and elevation angles to a given point), string it out on the map, estimate a distance, estimate a location—first? The smoke was at a considerable distance, and the landscape in that direction was bland: flat muskeg. I suddenly longed for a neighbouring tower to be closer, so they could assist me with a cross shot, but no one was close to me at Connaught. So I estimated 25 kilometres, figured out the longitude and latitude as well as the LSD (Alberta "legal subdivision" measurements), and called it in. And I waited. When would I hear a helicopter coming? I continued to wait. Around 40 minutes later the radio-room dispatcher called wondering how my smoke was doing. I reported that it was still visible and growing. Within a few minutes, a helicopter passed the tower on its way toward the column of smoke. After the initial assessment, the office decided not to action the fire, to let it burn out on its own. I watched the smoke grow in size, watched the storms continue to move around the hill and tried to breathe to help make the adrenaline subside.

Then I noticed something just at the bottom of the hill to my northwest—another smoke! Another smoke report, so simple at this distance. Getting all my information together quickly, I called it in, estimating the distance at 10 kilometres from my tower. It was

One of our first fires at Connaught, about 15 kilometres north of the tower.

fascinating to watch: orange flames licking out from the smoke, new flames and smoke bursting forth as it congratulated itself for finding new fuel. Again, about a half-hour later, the office called to ask me how it was progressing. I assured them that this new smoke was still visible, and soon the helicopter came into view once again. It completed several circles around the fire to do the initial assessment, then made its way to the pond just a short flight to the west, landing first to attach the bucket then, once airborne again, dropping low to fill the bucket from the pond and carrying it back to the fire to make the drop. Thus it continued, the only piece of equipment to fight that fire, making its way back and forth from the pond to the fire. In less than an hour there was just the odd bit of smoke remaining. He must have been getting low on fuel because he left, then returned again an hour later to finish the task. By the time he finished, there wasn't any smoke left to be seen. The burn area was so small that I could barely find it when I scanned the area later.

How fascinating it was to watch the helicopter at work. I had watched as crews trained while I was in Hinton and had read about the different courses on the bulletin boards—now it was all starting to come together; I began to get a feel for my role.

I was up my tower for the last time at 21:00, at which time there was still a lot of smoke rising from the first fire. I estimated it was about one kilometre wide, but I didn't think it had much depth, because I could see through the smoke at most points. (Viewing from a distance only gives you a one-dimensional perspective, so total area is hard to estimate.) I was learning so much from the opportunity to watch the fires burn. Although our training had touched on fire behaviour, actually seeing wildfires first hand brought the theories into focus. With prime burning time in the mid-afternoon when the sun is at its highest and humidity lowest, the evening now saw the fire grow drowsy. The cool temperatures and rising humidity reduced the burn rate and held the smoke closer to the ground.

During that evening observation, a silvery-white wisp appeared among the trees less than a kilometre away. I couldn't imagine that there was a fire atop the hill, because I hadn't seen lightning strike that close. As I studied that wisp in the evening light, smoke it was not. It was the northern harrier, come back to join us on earth. Dodging and darting silently among the trees, he was back from his adventures atop the storm clouds.

The crowning touch to that day of soaring hawks and the testing of new skills was when I checked with the fire centre the following day, and found that the GPS coordinates given by the officer sent to do the initial assessments confirmed that my estimated locations were, in fact, accurate.

I soared almost as high as that harrier. Perhaps I would eventually make a fire-tower observer after all.

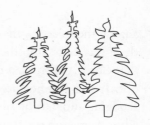

Birdsong and Lost Friends

My wayward container of hobbies and books finally arrived at Connaught the last week of June. I immediately installed the embroidery rack in the cupola and laid my knitting projects out in the cabin. My field guides were finally in my grasp. I proceeded to catch up on everything I hadn't been able to identify the past few weeks. It was a relief to be able to answer questions for myself again.

It was a wonderful feeling on a perfectly lovely afternoon to sit peacefully on the back step, watching the dogs sleep contently on the lawn while I knit, the forest and all its inhabitants going about their business as usual while I quietly listened and watched with busy hands.

June brought out the best in the birds. So many of the males were still singing their hearts out all day hoping for that special lady to come their way, and I was the grateful audience. The forest was never quiet, because they were in full voice long before I got up in the morning and well past bedtime. With patience and guidebooks at my side, I was able to decide that my ecosystem included phoebes (heard but not seen), Canada Jays (or grey jay or whisky-jack, if you prefer), rose-breasted grosbeaks, Tennessee warblers and both chipping and crowned sparrows. The tree and barn swallows had now arrived, and I loved watching as they soared around the yard. One poor barn swallow mistakenly flew through my kitchen one afternoon, but found its way back out with only a minor headache for its troubles.

The dogs and I scared the bejeebers out of a poor ruffed grouse one morning on our walk, but once she settled onto a safe spot on a branch, we spent quite a while checking each other out. Just as it is with Alberta northern flickers, these grouse are so much larger than Ontario varieties!

A personal favourite was a certain yellow-rumped warbler who was the most curious creature, spending much of his time close to the cabin. One evening, as I sat on the back stoop, he was sitting on the edge of the eavestrough peering down at me. When he became bored with that, he flew over to the tower, landing less than a foot above the sleeping dogs and investigated them for a while.

Reminiscent of the scene from *The Wizard of Oz* with Dorothy and her house caught inside the tornado and Dorothy looking at objects flying by her bedroom window, so it was for me when a raven hung on a breeze not five metres outside the cupola window one day. Although I tried to strike up a conversation, it found me most likely boring, and set its wings against the wind and went on its merry way.

The dogs took advantage of their wild surroundings. Ted left his city ways behind and became the intrepid critter hunter. There were "critters" (mice and voles) under everything: the cabin (outside, not inside), the generator shed, the stoops and the little garden between the tower and the cabin. His afternoons were spent moving from spot to spot, striking a frozen pose (not quite a "point") ears straight up, his head tipping classically from side to side as he listened to them scurry around. The funniest part would be when he would lie on the ground, place his chin on his front paws, and wag his tail at them, ever so much looking as though he was pleading with them to come out and be his friends. Ddreena would occasionally wonder what he was doing and have a sniff or two herself, but she couldn't understand the game.

One warm and pleasant morning when I took my coffee out to enjoy the sunshine from the front stoop, I was joined by a wee mouse in what seemed a very Beatrix Potter moment. It sat ever so relaxed just an inch under my foot (the stoops were of steel mesh easily seen through) while enjoying its own breakfast of sweet spring sprouts, occasionally stopping to preen its whiskers. This little episode, together with the excitement of the smokes of the past few days, brought a feeling of contentment. Peace and purpose woven together with the space to breathe, a job to do, the feeling of being an integral part of a working unit. Eyes with which to enjoy the beauty around me, a clean, warm and dry little cabin of my own, a blue sky above, green forests all around. My dogs peacefully asleep on the lawn before me, a mouse quietly grazing beneath my foot, my hands wrapped around a warm coffee mug.

Reminding me of clowns or jesters with their bright fuchsia tufted hats, a cluster of calypso (or fairy slipper) orchids bloomed in the arboretum. The western Canada violets bloomed at that time as well. Their flowers were far tinier than a dime and hugged the ground tightly. Flitting around like iridescent petals on the wind were the "blues"; the small pale periwinkle-lavender butterflies had recently arrived as well.

The fire north of me that I called in extinguished itself the next day. Since the area around Connaught was a "limited action zone," decision makers at the fire centre decided not to send crews and resources to it. That next day the wind changed every few hours, pushing the fire east, then west, then finally in a circle where it blew back over itself. Overnight it rained, and that was the end of it.

It had been a day of new and interesting thoughts as I sat in the cupola and watched that fire burn. My experience as a veterinary technician had encouraged me to lose the so-called

"Bambi Syndrome," and yet I had found myself at different moments throughout the day thinking of young animals who might not have been able to flee the fire, or of eggs just recently laid in nests and slow-moving species such as porcupines. Reality told me that forest fires are a natural part of the ecosystem, that regular fires reduced the amount of fuel (i.e., deadwood) in the forest and it was all for the greater good. My city side lamented all the smoke that fouled the air. One of the forestry officers from Jordan later explained the importance of a discontinuous forest and how small fires create natural firebreaks that help prevent major fires. So many perspectives to weigh, so many preconceived notions to lay to rest. The scientist in me told me I didn't have enough data to statistically prove my point. I stopped arguing with myself. This was not a man-made fire, this was Mother Nature looking after her earth as she had done for centuries before humans ever laid eyes on this land. If she had managed to create all the beauty below me by her skilled hand, then I deferred to her experience and wisdom.

The following night saw more lightning, so the following day I fully expected more starts, because the day had the perfect conditions for smouldering strikes to evolve into fires. Fires flare up in early afternoon, and true to form, I had a fire come up to my north once again, but far closer than the first one. Then late in the afternoon, yet another came up far to the northwest. After fretting over an estimated location, I decided that it was over the border in British Columbia. (We advise B.C. Forestry about them, but don't write up any sort of report.) Once again they decided not to action the fire to my north. I watched as it grew, casting long plumes of smoke that drifted east, sometimes emitting orange flames, other times seeming to struggle to remain alive. It rained that night, and when I looked the following morning, little smoke remained. A helicopter stopped by that morning, bringing a forestry officer and pilot to share a cup of coffee with me. We discussed the hows and whys of forest fires, whether to burn or to extinguish, and with his patience he answered some questions for me before taking off again to check on both fires. As I watched the helicopter make its way to the northern fire, I heard him radio in the report that the fire had now reached 50 hectares, all that accomplished in about 12 hours! My last observation of the day found only one small puff rising from the edge.

My 22nd day passed at Connaught. Twenty-two days without ever locking my doors, having keys or a wallet in my pocket, driving a car, flushing a toilet, watching TV, or indulging in ice cream, turning on an incandescent light or seeing a night sky or stars, so short was the northern night.

The helicopter had brought my mail, and it was wonderful to receive a letter from Steve. He told me of the things that he'd seen and done over his time here, the creatures that had wandered through, his memories of muskeg autumns. He had made arrangements many years ago to stay on longer than the official closing date, buying himself an extra month, and in turn had paid his own way out. I wouldn't be so lucky as to see those colours—I would be gone long before that.

Valerie called around 19:30 one evening to let me know that she was going to try and hike down to a lake not far from her cabin. She was such a responsible and mature young woman, telling me all the things she had packed: warm clothing, bear bangers, an air horn, flagging tape, food and water, etc. She figured it should take her about two to three hours to do the round trip.

The evening wore on. I tried calling her several times with no answer. I tried not to worry too much, because it never got truly dark; I was confident she would always have enough light to get home. Val was an avid reader, and I convinced myself she had taken a book and gotten involved with that down at the lake. I kept the phone beside me. It was my decision not to call the office when her estimated return time came and went.

At midnight, it was the darkest I had ever seen it. Obviously a storm was covering the western skies. Now I was worried. Another call produced nothing. I called Arlene at Redmond Tower (west of Stoney) and had her radio Val to be sure that it wasn't just a radio problem, and then decided that I was going to have to call the Duty Officer.

The Duty Officer occupies the "hot seat," a position filled on a rotating basis. The DO is at the helm of the fire centre; when a smoke is called in, he or she decides what will be done by whom and what resources will be sent out to the fire or incident. The Duty Officer is also on 24-hour call during that time.

The melodrama began with me trying to figure out what number I should call. There were too many choices without any indication of where to call during an emergency. I finally reached the Duty Officer. Poor fellow, he'd only been in this district for four days and had no idea where Stoney Tower was even located. But he seemed level headed and responsible. Even if he didn't know the lay of the land, he knew procedure. He asked me to continue radioing every 15 minutes while he made some other phone calls. He called me again around 01:45 to let me know that a helicopter would be ready at 05:30 to begin searching for her, and that I could stop radioing. When I asked why a helicopter couldn't be dispatched immediately, the reply was that helicopters don't fly at night. What a wonderful time to find out.

IN THE MORNING!!! Oh my gosh, poor Valerie! I didn't get a minute's sleep that night. I repeated to myself that she was responsible enough to be able to look after herself; it helped to quiet the yelling in my brain that I should have started a search earlier. The bulk of my worry was directed at variables that might be out of Valerie's control, such as bears (remembering that she'd already had one come knocking on her cabin door).

I was staring at my clock when 05:30 finally arrived. At 06:30 I called the radio room, but the operators had not yet received word, just that the helicopter had landed at Stoney. At 07:00 they were searching the area by air, confirming my fears that they hadn't found Valerie in her cabin. My worries increased again, but at 07:30 the radio operators called to let me know that the searchers had heard the sound of her air horn, and that she'd been seen walking, a good indication that she'd fared well through the night.

So Val missed morning weather reports, but radioed me not long afterwards with a cheery, "Hey! Guess what! I got lost!"

She had her weather report into the office 15 minutes later.

As we weren't allowed to use our cellphones for personal use between 07:00 and 19:00, I couldn't chat with her until the evening. She shared that except for a brief frustrating moment she'd handled herself quite well. She had accepted the fact that she was lost (she wasn't too far away from where she thought she should have been). She had forgotten to put that last piece of flagging tape on a tree after a nesting grouse scared the heck out of her as she arrived at the lake's edge, and she couldn't locate that spot again. She made herself a comfortable bed of leaves and spruce boughs and settled in for the night. She did have a bear (it didn't like the air horn) and moose pass by, and had quite the list of birds to look up, declaring the woods at night to be actually a very noisy and busy place. Except for the lack of sleep and some blood lost to the mosquitoes, she was fine. She had filled her time with yoga when she finally wasn't able to rest any longer. She received quite the hug from our supervisor, Bob, when he found her.

My mind swam with "what ifs?" and "should I haves?" I wasn't at all happy with how my role in the episode had played out, and I was determined to take what I had learned and to do better on what would hopefully not be a next time. I thought about all the times I had been hiking or kayaking solo, leaving my itinerary with my mother and assorted friends in case of emergency. Last night it had been my turn to decide if there was an actual emergency or not. Had I been able to balance that critical moment between giving Valerie her freedom and becoming concerned when she was overdue? The weight of responsibility of being an emergency contact was now much clearer in my mind.

Valerie promised me she would always call and let me know when she set her mind to wander about. It was also a confirmation of her faith in me as friend and neighbour, of which I still wasn't sure I was worthy.

Thank goodness my hair was already grey before this happened— no one would notice the difference at the end of the season.

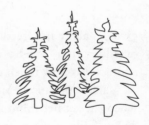

From the Ides to Solstice and Beyond

I found out what those chickens in the glass-doored rotisseries at the grocery store feel like as summer's arrival at Connaught became imminent. Although Valerie had been complaining for more than a week about how hot her cabin was, it took that many more days for the heat to creep north to my site, culminating with a high of 23.5°C. With a dry hot wind that gusted to 50+ k.p.h., I thought I would end up slow roasted in that cupola.

It hadn't rained for over a week, and what with those winds, I could now walk just about anywhere on-site without needing boots. The mosses and lichens withered. The cladonia (reindeer moss) turned to powder if I clenched it in my hand, the arboretum dried to the point of cracking. Flowers were blooming everywhere, including the Labrador tea. Every morning there seemed to be a new blossom waiting for me. Not to be left out, the insect populations bloomed as well. The smaller mosquitos, along with gigantic horseflies and blackflies, all clouded our evenings.

The fire indices had risen to just one number away from pushing me from high hazard level to extreme and an 11-hour day in the cupola. Many of the other towers were already at the extreme level, meaning long days for everyone. Several of the towers were also calling crossover conditions (when the temperature is higher than

Canadian Weather Fire Index System
(computed from weather data)

Moisture Codes

Fine Fuel Moisture Code (FFMC): Measures the moisture conditions of surface fuels.

Duff Moisture Code (DMC): Measures the moisture conditions of the 2–4 inches (5–10 cm) below the surface.

Drought Code (DC): Measures the moisture conditions of the deep duff 4–8 inches (10–20 cm) below the surface.

Fire Behaviour Indices

Initial Spread Index (ISI): Given dryness and wind speed, gives an indication of how quickly a fire will spread. (Blends FFMC and wind speed.)

Build-up Index (BUI): An indicator of the amount of fuel dry enough to aid combustion. (Blends DMC and DC.)

Fire Weather Index (FWI): An indicator of fire intensity. (Blends ISI and BUI.)

the relative humidity), always an anxious time for the tower community and the fire centres.

At that time there were fires being fought in our district. An east wind had blown the smoke from these fires up to Connaught, reducing the visibility to under 25 kilometres for most of the day. Signant and Misty Mountain, the towers closest to the fires (the Talbot Lake Fire, Tall Creek Fire, and a third somewhere south of Wood Buffalo Park, near the Caribou Mountains), both reported visibility of only 5 kilometres. I had heard that one of the fires went from 100 to 400 hectares in one afternoon. The House River Fire, southeast of Fort McMurray, had surpassed 250,000 hectares. And that was the news from only two of the districts in the province.

All of this, of course, was happening just around the time of my first service. Service day (Rick at Signant dubbed it "Christmas") was the once-a-month arrival of groceries, supplies and mail. With the fire centre staff and helicopters heading out to fight these fires (60 helicopters and 1,000 firefighters at the House River Fire), it seemed likely service day was going to be put off. I had started working on my grocery list. In this district, the towerpeople phoned in their grocery orders directly to the store. I hadn't realized it was going to be such a struggle, and I began to understand how much of a visual shopper I was. Just staring at a blank piece of paper wasn't working for me. I needed to see jars, labels, quantities. I had adjusted my order three times by the time I told myself I wasn't allowed to call the grocery store anymore, because I had heard the exasperation in the clerk's voice the last time I called. Requests for non-grocery supplies took an interesting turn when the only fellow in the warehouse with authority to purchase without getting multiple signatures had been shipped out to work at a fire. Also, every office person who knew the combination to the office safe where our "slush money" (personal cash left in the office at the beginning of the season for non-grocery store items) had been "exported" as well.

Just prior to this, helicopters had come in three times that week for a variety of reasons, including delivering propane and repairs. As the last chopper was rising from my yard with the inevitable tornado blowing the forest, the dogs and me around, I couldn't help but think I had had my fill of humans and helicopters that week. I was looking forward to some solitude again.

I surprised myself with that feeling. Friends and family had been asking me questions such as, "What will you do with all that time?" or "Won't you be lonely?" The answer was a resounding "No!" The time was flying by! I hadn't been getting to bed until well past midnight (it's surprising how difficult it is to judge bedtime when the sun is still above the horizon), and despite being up at 06:00, I still couldn't get caught up with the chores.

The peas in the garden were beginning to look like real pea plants. I had taken the Swiss chard seedlings out of their egg-crate hatchery and added them to the little garden with the leeks and spinach. A second species of orchid, pale coralroot, was now blooming behind the tower. Bearberry and the lowbush cranberries were in bloom, the tall lungwort had added its rich blue to the myriad colours around the site, the fireweed was growing in leaps and bounds. Wild chives flourished where one of the tower guy wires met the ground. Small orange butterflies joined the lavender ones; large white butterflies and swallowtails floated across the yard each afternoon. The end of my cabin now held a barn swallow's mud nest.

At different moments during the day I would see a red squirrel entertain himself by seeing how close he could run by the dogs while they were sleeping, or listen to my warbler friend who continued to watch me daily from his perch on the tower ladder. He would sing from morning until night, and I swear I could tell his voice from among the hundreds coming from the woods. Sometimes we had long talks about all this singing and hoping for someone special to come along. I tried to tell him that he needed to think in a more independent way, that he just couldn't spend all this time singing and forget all the other parts of his life. But he sang through all my words of wisdom, finally turning his yellow back to me. I hoped that he would find someone who appreciated that kind of determination.

The winds brought me my first experience watching a coniferous pollen blow. It had been described to us during training as one of a variety of "smoky" events that might be hard to discern from genuine smoke. It was wispily beautiful. The trees would begin to rustle, and large dusty gold clouds would rise from the forest and drift away in waves with the wind.

I had also been getting phone calls from the British Columbia forestry fellows, because they were having computer problems and appreciated being able to check in with me for a visual update and for weather data, especially when their radar showed lightning strikes on their side of the border. They were fun to chat with; a cheery bunch they seemed, and I welcomed that residual good feeling one has whenever one is being extra useful.

The dry heat finally turned to rain. A grand total of almost 12 millimetres fell over three days. If that much had made it up the hill to me, I surmised, much more had fallen on the plains below, with the pattern continuing of very few storms bothering to expend the energy requisite to climb up to Connaught. The heat washed away with the rain, returning our mornings to a balmy 3°C. (Calgary was very hot at this time, but Waterton had so much snow they were considering calling a state of emergency!)

My indices fell with the precipitation, giving me back a bit of time on the ground. Although enjoyable, it threw my routine out of whack. Several of us had been discussing it on the phone. You would just get into a groove, go up to the cupola, spend the whole day up there, then it would rain, and down you go, back on the ground not remembering what job you were working on last. The day that my groceries arrived, I lost track of time entirely. The helicopter arrived in the morning, and by the time I got everything put away and indulged in some of my mail while still making my regular runs up the tower, it was after 18:00.

The new chilly temperatures kept the insects at bay, and the dogs and I could once again go outside for an entire day. We would spend most of the morning "snipping" the lawn, then after lunch spend a bit of time sitting out on the stoop knitting. It was lovely. With the wind out of the north, the hazy-smoky skies brought to me by the southern winds were gone, and I could once again see the B.C. and N.W.T. landscapes set in motion by cloud shadows galloping across them.

My fresh groceries allowed for a few more menu choices and treats. A set of sheets that actually fit my bed was the highlight of the packages from home. Also included were some current newspapers to read, and adding even more sunshine to my day was an assortment of other treats from thoughtful friends. Even the string that tied one of the packages was cause for celebration, because it was perfect for tying the peas to my recently constructed trellis.

I had been looking forward to solstice. Partly because I wanted to see just how long that day was going to be; partly to celebrate in a natural way all that my soul had been recently given.

All kinds of colours and textures continued to feed my senses as summer flourished. A third species of orchid had joined us: the round-leaved orchid. I never imagined so many species of what I had previously imagined to be rare out here on the muskeg. These little polka-dotted beauties were ever so tiny and occupied the northeast section of the lawn. The summer meadow effect was now in full swing as the yarrow and fireweed were in bud; the arboretum was a sea of Labrador tea blooms, everything growing from a thick carpet of bunchberry. The soft pinks of the wild roses charmed these days, almost as though they purposefully planned to herald the arrival of summer. The tiny little bog cranberry plants ended each branch with even tinier little pink blooms as a border along the edge of my vegetable garden. The blue-green tips of the spruces and pines were beginning to extend their new needles.

The forest had become quiet, the male birds no longer called for mates. Other than morning songs and evening vespers, there was very little birdsong anymore, neither were there many birds moving around. Undoubtedly everyone was on their nests. Birds populate out here in differing ways from birds in the city. Although there were species that had larger seasonal populations, such as the crowned sparrows, chipping sparrows and warblers, other species like ravens and whisky-jacks seemed to only have one pair inhabit an area of the hill. I had only intermittent visits from the swallows now, and my little warbler friend had either found himself a girlfriend or decided to move along to greener pastures, because I hadn't seen him in almost a week. But my bunnies still arrived every evening, munching on the various delicacies of the lawn.

Visitors arrived from the air once again when the office finally sent someone to level the cabin. Connaught had a long history of leaning. Given the muskeg, it wasn't hard to understand. The comms (communications) tech who had earlier come out and fixed my radios had noted that the cabin used to lean to the north, but was now sinking even lower to the west. The past couple of weeks the cabin had sunk so low that the back door would only open a foot, and the bottom door trim was beginning to catch on the step, bending it upwards. The carpenter arrived with huge jacks in tow, and within a couple of hours everything was on the level again.

What a difference! (The carpenter had declared the cabin to be on a one-over-four slope!) I had gotten used to walking on an angle, always looking for Ted's ball somewhere along the west wall. Now I didn't have to keep my heels dug in under the desk and my elbows planted on the desktop for fear of rolling away! Not only did the door open freely, but I had to add another 30 centimetres of downspout so the rainwater could empty directly into the barrel again. I had almost shortened that piece because it was buckling and gaping so badly as it pressed down upon the top of the water barrel.

Smokes were called in by the southern towers in our district, and now from early morning the radios chattered, sending instructions to bird dogs (small planes that lead the bombers) and letting forestry officials know where crews were being dropped off. Harold, who manned Basset Tower, was asked to start reporting the weather every hour. Tom, from Zephyr Tower, told me that a fire farther south that had started a few days earlier near Preston was out of control. All crews had been called off, and it was being allowed to burn. Due to the winds that day, the fire had started to crown, advancing at a rate of 46 metres a minute. You can try your hardest, but sometimes Mother Nature still wins.

On the fun side, the towerfolk in this district were starting to get a bit silly during evening scheds (I had begun to worry that it would never happen). One particular night, Val, Arlene, Cathy and I got together and did a little rewrite of the old Carpenters song, "Top of the World," each of us singing one line:

We're on the top of the world looking down on creation,
And the only explanation we can find,
Is cuz the hazard is high and we're stuck here by and by,
And that's put us at the top of the world.

We got Harold to introduce us so the office would know who was involved, because, in order to keep the song moving, we wouldn't be doing our standard QRU and station identification. He did a marvellous job, including an enthusiastic, "Take it away girls!"

And thus the dogs and I settled into our tower life. Sometimes there was excitement; other times there were weeks of only peace, beauty, new sights and sounds, the motion of Mother Nature moving quietly but steadily by us, but no great adventures of which to speak.

Ted's prowess as a fledgling hunter improved with time at Connaught. After a month or more of singing to our rodent friends to entice them to come and play with him, he finally made a kill. We had been walking around the arboretum when he dashed off to chase something. Ah well, I thought, the critter has nothing to worry about. Next thing I knew, I heard wild squeaking, and dashed back to see what I could do. Just as I reached Ted, he swallowed the poor little thing, still squeaking as it slid down his throat. He spent the next 24 hours curled up in a little ball under the desk in the cabin, I presume from a stomach ache, which I reminded him several times he deserved.

Several days later, I heard him snarling just outside the front door. When I went to see what the fuss was about, he had yet another dead rodent at the bottom of the stairs. Ddreena had come to investigate, and Ted was insistent that he wasn't about to share. I went out, sat on my heels beside him, and inquired if he would like me to dispose of it for him. He replied by picking it up and swallowing it whole (at least it was dead this time) and smacking his lips. Less than an hour later there was another carcass at the front step, but that one was quickly forgotten as he ran back out to find yet another victim. That one went out to the grass clipping pile. As I laid that poor little saliva-covered body out on the clippings, I couldn't help but think how soon we can change from hunting for sustainability (imagined or real) to being wasteful.

For all the preparations, worries and stress of getting pets out to a tower, I couldn't imagine not having them at my side. But pets at a tower are a huge responsibility. Forestry will not send helicopters in cases of emergency; the towerperson must be prepared to be both doctor and nurse and to have the best possible veterinary first-aid kit in tow. When the fire centre and warehouse are located in small towns, particular brands of pet food may not be available, and one must bring the entire season's worth of food or make arrangements for it to be shipped in as needed.

I have a wonderful veterinarian who allows me to take a good selection of remedies such as antibiotics and ointments. If something should come up, all I have to do is call him, explain the situation, and follow through on his suggested course of action, meds in hand. Then I simply request a resupply of those meds. A former career as a veterinary technician has been a great advantage.

As I unpacked from our second service day on Connaught, the first thing I looked for was the box that would contain Ddreena's meds, a mild anti-inflammatory to help her move around better. (I had originally brought enough for the whole summer, but I ended up having to increase her dosage, which had left me short.) However, they were not to be found; poor Ddreenie, she just didn't deserve that. We tweaked the dosage again, kept the aspirin handy, and made it through to the end of the season.

There was never idle time for fire-tower observers. Rainy days were filled with indoor chores that never seemed to get done on sunny days when we spent long hours in the cupola; sunny days saw us trying to get chores like laundry or site maintenance done around our responsibilities "upstairs." As I had no days off, there was very little chance of drifting too far away from the cabin; schedules needed to be adhered to. Embroideries were completed and new ones started; sweaters destined as Christmas gifts were packed away. As a volunteer with Yukon Learn, I corresponded regularly via the Internet with a student as we worked together to improve his English.

Friendships developed among the towers. There was a real sense of keeping an eye out for the towers that surrounded you; a sense of camaraderie. Between evening phone calls and the regularly scheduled radio transmissions, there remained a human/societal element to our lives. Sunday evenings saw us arrange goofy "ice-breaker" type of exchanges for evening scheds, such as sharing what books we had been reading (everything from Ivanhoe to Dickens; from genetics to romance) and what music we had brought. Sometimes we got creative, coming up with silly things like advertisements (topics left to our imaginations), and occasionally the radio operator at the fire centre would join in. In this district, the towers put out an annual newsletter, and I was made editor, since I had the only computer and could send the final file into the office for printing.

Time heavy on my hands? Never.

Hello, We Just Stopped by to Bug You

I've never quite understood people who can't accept insects. I suspect these are the same people who think they can change parts of an ecosystem without impacting the rest of it. No creatures—vertebrate or invertebrate—would be here unless they were successful at filling a niche. The first question most people ask when I say I've spent a summer on the muskeg is how did I manage against the biting flies. I never expected them not to be there, so I went prepared for them. With the use of head nets, bug jackets and repellents, I accumulated only about three bites all season at Connaught.

The beautiful hot early July weather inspired the worst part of the insect explosion, and everything was biting at the same time. Ddreena stubbornly insisted on spending part of the day outside even though Ted had already given up and decided he was better off in the cabin. I had only been up in the tower for about an hour one morning when Ddreena started bellowing and staring at the back door. When I came down to let her into the cabin, I saw that the horseflies had feasted on her feet and lower legs, and with all that anticoagulant, she dripped blood everywhere.

Ted's problems were the blackflies. They managed to find his belly under all his long hair, covering it with ugly purple hickies. Their abdomens bloated with blood, they would hitch a ride back into the cabin with Ted, then fly off (if they could even manage to become airborne) to the windows. If I smacked them, there was blood everywhere. If I used the spray to kill them, there would be spray residue all over the windows. There was just no winning. I cleaned windows weekly to minimize the "slaughterhouse splash" effect. I tried fruitlessly to get Ddreena to wear the head net, as I had a bug jacket to help me survive the time I spent gardening or cutting the lawn, but she just wouldn't have any part of it. The best I could do for the dogs was apply bug spray, trying my best to only wet areas they couldn't lick, and adding pre-sprayed bandanas around their necks. When it came time for the three of us to come into the cabin, we would run the last 30 metres to shake loose any hitchhikers.

My little warbler friends joined us again with the insect bloom, but now it was to forage rather than sing for their mates. I would watch them from the cupola as they took their fill of horseflies. Hundreds of flies spent each afternoon swarming the warm side of the cabin. From inside the cabin I could hear the constant hum of horseflies and the tapping sounds as they repeatedly hit the windows from late morning until dusk. The warblers would perch on the eavestrough, almost standing on their heads to look between their toes to see underneath the overhang. A short dive with a half-twist and they would have a fly in their mouths, then land back up on the roof where they would repeatedly hurl it onto the shingles to stun it. It amazed me that there were only two or three birds at any given time that did this. With so many warblers in the area, I would have thought that there would be line-ups for such an easy meal.

But the unwelcome insects were in balance with those that brought only beauty with them—the butterflies. There were swallowtails, white admirals and sulphurs, as well as many blues, skippers and checkerspots that I never managed to photograph.

Shadowing the biting flies' arrival were the dragonflies. I enjoyed watching them catch and eat mosquitoes, zigzagging around the yard even though they hardly had to move for the next morsel. It was noticeable how quickly they grew in size over those weeks. I cheered for them from the windows, calling out for them to bring their friends along to the feast.

Thus another of Mother Nature's paradoxes was presented to us: the warm weather was lovely, but it meant less enjoyable time outside, whereas cold blustery weather allowed us better-quality time outdoors, since the biting flies weren't out or just couldn't stop fast enough as they blew past us. I had heard other towerpeople say that Steve used to go out to his garden every evening right after scheds. I could never understand how he managed that, because the flies were at their worst in the evenings. He must have been one tough old soldier.

The Opening Act Closes with Curtains of Shimmering Green

A late-night phone call during one of my final nights at Connaught found me gazing out my bedroom window at Venus twinkling at me as I made my way back to bed—the first star I'd seen since leaving Calgary. I looked around at the boxes stacked everywhere as I prepared for the helicopter to come and whisk us away.

The impact of that call had my soul lamenting the missed opportunity to sit quietly in awe beneath stars and northern lights before I left this north country. It just wasn't fair that they were closing my tower two weeks early.

Back in Jordan, Mother Nature smiled upon me one last time during the several days I spent in the bunkhouse before heading home. I had left the group from the office behind at one of the local bars. That evening was my last night, and because of the long drive home the next day, I thought it best to get a full night's sleep.

As I rounded the final curve in the road before arriving back at the bunkhouse, what was awaiting me but an aurora borealis.

The dogs and I stood in the parking lot as it shone all around us. Sheer green curtains waved in imaginary breezes, watercolour washes drizzled down the midnight-blue canvas finishing in spectacular

pinks; and whorls turned inward upon themselves and then unrav-
elled again before fading away.

The heavens that shone above us, the stars slowly tracking across
the sky, showing the gentle turn of the earth and the change in the
seasons, were bidding us farewell. Had they heard my laments?
Was this little corner of the earth such a happy place that it celebrated its good fortune with its own private fireworks? Oh, that I
was so lucky as to have been part of that space, that energy for that
glorious summer.

I wiped away my tears, and bid a final farewell to the heavens.
Even with the abrupt end to our tower season, I had managed to
experience what I had hoped for: the weave of the ecosystem of
Connaught, the midnight sun, the return of stars and, finally, the
northern lights.

"This is XMR 426, down for the season."

**Connaught Tower
Incomplete Species List
121-10-W6M
Elevation 780 metres
May 15–August 1, 2002**

Aspen

Asters, not yet in bloom
Asteraceae spp.

Bearberry
Arctostaphylos uva-ursi

Birch

Black spruce

Blueberry
Vaccinium spp.

Bunchberry
Cornus canadensis

Calypso orchid
Calypso bulbosa

Common nettle
Urtica dioica

Common pink wintergreen
Pyrola asarifolia

Common yarrow
Achillea millefolium

Cranberry
Vaccinium spp.

Crowberry
Empetrum nigrum

Field horsetail
Equisetum arvense

Fireweed
Epilobium angustifolium

Graceful cinquefoil
Potentilla gracilis

Green saxifrage
Chrysosplenium tetrandrum

Labrador tea
Ledum latifolium

Lodgepole pine

Mouse-ear chickweed
Cerastium arvense

Northern grass-of-Parnassus
Parnassia palustris

Northern green bog orchid
Platanthera hyperborean

Paintbrush
Castilleja miniata

Pale coralroot
Corallorhiza trifida

Pink pussytoes
Antennaria rosea

Reindeer moss
Cladonia spp.

Round-leaved orchid
Amerorchis rotundifolia

Stiff club moss
Lycopodium annotinum

Tall Jacob's ladder
Polemonium caeruleum

Tall lungwort
Mertensia paniculata

Western Canada violet
Viola canadensis

White spruce

Wild chives
Allium schoenoprasum

Wild rose
Rosa arvensis

Wild strawberry
Fragaria vesca

Willow

Yellow avens
Geum aleppicum

Whisky-jack Mountain Lookout

Upon the announcement of the early closure of Connaught, I had sent out emails to the other districts hoping to find late-season sub work. I was delighted to receive a call from Pineridge District asking if I might fill in at several of their towers. So there was only time for a quick drive back to Calgary to repack (during which we bid farewell to the ol' Toyota when the engine blew up near Preston, leaving us stranded on a long weekend). I had a quick haircut, enjoyed a meal with friends, left Ddreena with a sitter, and then Ted and I were off to Rumseytown.

Bunkhouses always offer surprising new acquaintances to share interesting ideas with. I had the opportunity to spend a quiet evening chatting with a young range biologist and discuss the work he was doing in the area. Bright and early the next morning we were on our way to our short four-day assignment on Whisky-jack Mountain Lookout, which sits at 2,128 metres above sea level.

Upon our arrival, we exchanged greetings with Patrick, who had been a classmate in Hinton, even if it was only for a moment while we swapped places in the helicopter.

What an entirely different environment a mountaintop is! The cabin was square (not octagonal as I had thought) with windows all around. The counters, cupboards, kitchen (except

for the refrigerator) and bed had been built below the window
levels to give a 360° view. There was a tidy wooden deck all around,
which Patrick had just finished painting before leaving for his days
off. The obligatory generator shed and biff (with a bucket below
the seat instead of a pit), two propane pigs a couple of metres away
and a large wooden helipad completed the Spartan setting. After
spending all summer at Connaught's altitude of 780 metres, I was
a bit drowsy at my regular mid-afternoon low-energy time, but
otherwise seemed to manage in the thinner air.

The first day welcomed us with warm sun, strong winds and big
puffy clouds that threw racing shadows across the landscape so very
far below us. To the south and west were mountains as far as the
eye could see, many snow capped, looking ever so much like people
in a theatre, their heads filling the spaces between the heads of the
adjacent rows. The frontal range, the Brill Mountains, gave way to
foothills to the north. A ribbon of highway landmarked that transition; to the west, the Victoria River flowed north through The Gate
with the Brill Reservoir a mere spot on the horizon.

The peak on which the cabin sat is one of a series of ridges. Because
the site is above treeline, there was only loose talus underfoot with
some pillow-sized clumps of grass. Just to the east on a neighbouring
ridge sat a cabin and pen that belonged to a group of bighorn-sheep
researchers working on a long-term study of the local herd. On a
mountaintop to the west I could see Stenhope Lookout, my next
destination. I tried my best to find Helios Lookout in the distance,
where I would be going in just a few weeks' time, but I was never
sure if I found the right peak. I was happy to see that there were
plenty of places to walk, because I felt cabin fever set in not long after
arriving. Ted and I indulged in a lovely evening walk on that first
night and found quite a few paths for future meanderings.

The local wildlife included two golden-mantled ground squirrels
and one red squirrel that enjoyed running the railing around the
deck, giving Ted a good workout. Just outside the door a little stone

From its location on the easternmost range of the Rockies, the view from Whisky-jack Mountain Lookout beyond the cabin to the east seems endless.

wall had been built, where the golden-mantled ground squirrels hid and which protected the tiny little carpet of lawn someone had sown. Patrick had found rocks that reminded him of houses and had painted them as such, creating a little village at the bottom of the wall.

After being so isolated at Connaught, I was somewhat surprised to look up late that afternoon and spot a hiker sitting at the far edge of a long, gently sloping extension of the peak that extended to the north. She didn't come any nearer, and from what I saw of Whisky-jack's visitors' log I could expect few people to arrive.

With further studying, the flat landscape to the north revealed a number of flarestacks and two drilling rigs. I was able to see dust rising from traffic moving down gravel roads (another sight I hadn't experienced at Connaught), and I could see the lights of Rumseytown twinkle after dark.

Our first mountain sunset was beautiful, with the pink tones extending unbroken from the west through to the east. I settled in for the night on the single bunk in the northwest corner of the cabin. With no curtains over any of the windows, the heavens were open to my view. It was hard to get used to the early dusk this far south again after being in synchronicity with the near-midnight sun of Connaught.

The next day was our trial by weather. It was lovely until noon, so Ted and I headed out for a walk, collecting rocks, fossils and plants as I tried to make up a quick species list. The plants were composed of tight little clumps clinging to the talus, with many species related to familiar garden plants. I saw potentilla, sedum, cinquefoil, exquisite midnight-blue delphinium, and mauve and white asters or fleabanes, to name a few.

My introduction to the local fossils was the collection lined along the windowsills of the cabin. How interesting to be so high above the world, and yet see seashells, sea plants and corals pressed into the rocks and to imagine what events led to the earth pushing the sea bottom toward the sky. The peaks and layers of rock protruding through the surface at anywhere from 45° to 90° angles showed how the rocks had responded to forces from below. I didn't have to walk far to find fossil-encrusted rocks—they were everywhere, and I didn't hesitate to take a few choice ones with me. We surprised a hoary marmot and watched as it disappeared over a ridge while we walked to the closest tree, about 20 minutes away. The sparrow-like birds, I finally decided, were Sprague's pipit, but I couldn't identify some resident killdeer/sandpiper-like birds.

Just after lunch the sound of thunder reached us; clouds rolled in, nearby towers called in lightning strikes, and then I was fogged in. The wind rattled the cabin the rest of the day, encouraging me to turn the heat up more than once. It also rained, hailed and snowed seemingly in rotation the remainder of the day, changing the landscape to white with the appropriate part of the cycle.

Upon arriving at the loo, having had to brace myself to avoid the wind pushing me down along the way, I discovered quite the draft from under the seat; it was a worry that the pee was going to blow astray before it made it the few inches to the bucket! After walking around the outside to appease my curiosity, I found a hole on the west side of the outhouse that probably had once had a vent over it. Because the outhouse was double-walled, the wind slid down between the two layers to the concrete pad where it was forced under and back up in the area under the seat. A little ventilation can be helpful, but really!

From our haven in the cabin looking out along the ridge, it looked almost physics-defying to see the low clouds blow up and over the ridges, much like watching water run uphill only to rush down the other side. Like misty ballerinas, they swirled and curled, dancing with the winds before rushing off to new adventures.

The wind blew so hard that night that I awoke several times, not exactly sure if it was because of the sound or the cabin shaking. Eventually, morning dawned and the sky cleared, allowing Ted and me to set out on further walks and fossil searches. It wasn't a matter of finding them, but a matter of deciding which ones were worth taking with us. Since the path back to the cabin was all uphill, I had to be fussy about how much weight to carry. Poor Ted. As much as he wanted to go out on these walks, his feet had become sore because they weren't prepared for the move from mossy muskeg to this stony talus. Yet he remained stoic. Solo canine for the first time in his life, he learned to adapt to this mountain existence just as I did—he even learned to pee in gale force winds. (Not that he could lick his toe and hold it up to check wind direction, but he did learn to look around, checking to see which way his fur was blowing before making any rash decisions.) The red squirrel filled the spot created by Ddreena's absence and kept him entertained. It was quite comical to watch the squirrel call Ted over to the edge of the helipad, then slip underneath and come up on the far side, chattering and laughing maniacally that he had pulled such a good joke on a dog.

Later that day we wandered along the edge of the ridge to the southwest and found a little monument at the top of an adjacent peak. It was a stone cairn with a white post rising from the middle, with four names and the PPCLI (Princess Patricia's Canadian Light Infantry) logo carved into it. Patrick served with the regiment, and it was obvious that this was his handiwork. It wasn't until a later conversation with the sheep researcher that I realized the four names on the post were the four soldiers who had recently been killed by American "friendly fire" in Afghanistan. Since the PPCLI logo contains the image of a ram, and this mountain had its own resident herd, it seemed right that this private memorial stood in their honour on this particular mountain.

That evening saw the arrival of the Bighorn sheep. Ted and I walked to the east point to sit quietly and watch 29 of them grazing and scaling the slopes. I tried to read their collars as I peered at them through the binoculars. Later in the evening, 15 came just below the cabin and grazed until dark. The researcher had told me that the herd used to number around 120, but these days it stood at only around 33, with only one mature ram who had managed to survive the past winter. They hadn't been able to come to a conclusion on what pressures had caused the drop in herd size.

I set the alarm clock that last night so that I could watch the Perseid meteor shower, but when I awoke it was so cold and windy I couldn't muster the strength to sit outside. I did watch out the windows for a while, but with the roof obstructing a good part of the sky, I only managed a few sightings.

Our final day at Whisky-jack Mountain was filled with sun and winds that gusted to 85 k.p.h. I spent a few lovely hours on the lee side of the cabin peacefully knitting and soaking in the view. I called in one smoke and conferred with neighbouring towers on a couple of others. (After being somewhat independent at Connaught, it was a bit of an adjustment to remember that other towers might be able to see the same smoke. Since they were all far more familiar with the landscape, it was best to confer with them before bothering the fire centre.) Calling smokes in a new location always takes a bit of bravado to control the stage fright, so I was glad I got that behind me.

The following morning, the radio chatter told me that the helicopter was at Shandray staging camp and they were ready to bring Patrick back and whisk Ted and me away. The winds rose once again, and although they made it up the mountain, they couldn't make a safe landing on the helipad. In the outhouse, the winds were high enough to raise the toilet lid. I couldn't help but tease Patrick and our tower supervisor that we could devise a new wind scale and I asked if Patrick had ever taken the time to calibrate the height of the lid-to-wind speed.

It was several hours before the wind spirits gave their approval, but the helicopter eventually managed to land long enough for Patrick to leap out with his gear as I threw my luggage, fossils and Ted aboard, and we were off back down to Shandray, then off to Stenhope for another four days. Larry had been on the radio several times, mentioning how he was looking forward to getting these few days off, and I was counting my blessings for all these wonderful adventures.

"*This is XMU 59, clear.*"

Beaufort Wind Scale

Beaufort Scale	Description	Specifications	Wind Speed Range (km/hr)
0	Calm	Smoke rises vertically	<1
1	Light air	Wind direction shown by smoke drift but not vanes	1–6
2	Slight breeze	Wind felt on face, leaves rustle, vanes moved	7–12
3	Gentle breeze	Leaves, twigs in constant motion, wind extends light flag	13–18
4	Moderate breeze	Dust raised, loose paper, small branches are moved	19–26
5	Fresh breeze	Small trees in leaf begin to sway	27–35
6	Strong breeze	Large branches in motion, whistling in wires	36–44
7	Moderate gale	Whole trees in motion	45–55
8	Fresh gale	Twigs broken off trees, walking difficult	56–66
9	Strong gale	Slight damage to buildings occurs	67–77
10	Whole gale	Trees uprooted, considerable damage to buildings	68–90

Whisky-jack Mountain Lookout
Incomplete Species List
39-13 W5M
Elevation 2,128 metres
August 9–13, 2002

Alpine Bistort
Polygonum viviparum

Alpine chickweed
Cerastium beeringianum

Alpine goldenrod
Solidago multiradiata

Asters (white and mauve)
Asteraceae spp.

Common yarrow
Achillea millefolium

Dwarf saw-wort
Saussurea nuda

Fleabane (white)
Erigeron sp.

Grasses

Low larkspur
Delphinium bicolor

Lyall's ironplant
Haplopappus lyallii

Mountain gentian
Gentiana calycosa

Parry's townsendia
Townsendia parryi

Purple saxifrage
Saxifraga oppositifolia
(presumed—flowers not seen)

Purple vetch
Astragalus spp.

Rose-root stonecrop
Sedum integrifolium

Snow cinquefoil
Potentilla spp.

Sweet-flowered androsace
(rock jasmine)
Androsace chamaejasme

White camas
Zigadenus elegans

Yellow mountain avens
Dryas drummondii

Stenhope Lookout

Stenhope Lookout was very different from Connaught and Whisky-jack Mountain. What an experience, as a first-year towerperson, to be able to see and experience such a variety of towers in just one season. The sights, sounds, smells; the contemplations, the places my mind explored, the people I met, the stories I heard. I came to believe that if I were to travel another million miles in my lifetime, I would never come across Mother Nature in the same form twice. She is a morphological wonder.

Cam, our tower supervisor, and I drove immediately from Shandray staging camp to Stenhope upon our landing from Whisky-jack Mountain. In a rural sense, it was just down the road a bit, through a locked gate, and then up a few hundred metres, but navigable by vehicle. Larry was there waiting for us, eager to start his days off (not a regular part of a towerperson's schedule, this was a special treat and Larry had grasped the opportunity with gusto).

Stenhope Lookout sits atop Shandray Mountain, at an elevation of 2,083 metres. The environment was much different from Whisky-jack Mountain's bare rock ridges, despite the mere 50-metre difference in altitude between them. Sitting in the middle of a large meadow, with no talus at all, Ted and his feet were in heaven, and I had to admit the idea of a large level area held an attraction for me as well. Quite the assemblage of wildlife kept Larry company on that mountaintop; a few pairs of golden mantled squirrels, a sizeable collection of hoary marmots of every age imaginable, whitetail deer (as seen in Larry's photo album), bighorn sheep and foxes spent summers with him.

The cabin was older, quaint and the most cottage-like of the three I had worked in so far. The cupola was mounted on the tin roof, with the ladder into the cupola attached to the wall of the general living area (kitchen and living room) inside the cabin. Larry had made this space his own over the years. What this site had that made it different from many (but not all) was a telephone land line and electricity. Larry had taken full advantage of this and brought along his own television set, a cabinet full of videos and a stereo. He spent his spare time oil painting in his studio set-up in the little office, and finished pieces adorned the walls throughout the cabin.

But the view was what Stenhope was all about. Toward the southwest lies Watertonesque Lincoln Lake (for those who have had the good fortune to bask in the views of Waterton Lakes National Park). From atop Shandray Mountain, one looks out over a wide, conifer-covered rippled valley with many small bodies of water cupped gently among the hills, and follows the Victoria River to its origins where it flows over the dam that creates Lincoln Lake. One's gaze to the southern horizon is captured by the panorama of steep, snow-topped mountains that frame the lake itself and march on seemingly with no end.

Stenhope is one of the most frequently visited towers in Alberta, with over 2,000 visitors a year. During the four days I was there, at least 75 people stopped by. There were two groups from the hostel in Lindsay (Ted and I greeted travellers from Japan, Korea, Taiwan, South Africa, Australia and Texas), a group from a boys' camp, families, couples, singles; on Saturday alone I had around 15 quads (all-terrain vehicles) parked at the edge of the meadow. There was a log book for people to sign, and Larry allowed people into the cabin (it's a very odd feeling to suddenly have to worry about an appropriate time to bathe lest people start walking in the door) and let them climb up into the cupola. It's difficult to keep track of priorities when one has to answer questions (the most popular question being, "Where's Larry?") as well as watch for smokes. As for Larry, he loved the people. Families that arrived year after year would find that Larry remembered all the children's names and delighted in seeing how much taller they'd grown over the winter. It was well known that many school classes and groups such as Boy Scouts hiked up the mountain to have Larry tell them about life as a towerman and about wildfire management. Larry and Stenhope were made for each other, and this benefitted everyone: Larry, the public and Forestry.

All this continuous human commotion was disconcerting for me. My method of communicating with the environment had become internal and unspoken, listening to and watching what was around

Weather can be quite dynamic on a mountain lookout; the snow pictured here arrived on August 15.

me. Now with all this external stimulus I had a mild case of auditory overload. It was almost like I *couldn't* hear the earth any longer. The subtle bits of information from the wind, the birds, the clouds—they were all hidden behind this buzzy human hum.

Snow arrived and was quickly followed by rain, only to be followed the next morning with 38 millimetres of snow and a temperature of only -1.5°C (by lunch it was still 0°C)—and this was mid-August!

Saturday and Sunday, my last two days, were glorious. I spent almost all of Sunday sitting on the bench outside the cabin, handheld radio at my side, knitting and taking in the view of Lincoln Lake, answering visitors' questions. Monday morning saw Larry arrive back at the lookout, well rested and saying how much he'd enjoyed his few days off. With no missed fires to report, I left him on his mountain ready to take the helm once again with a smile on his face, as more quads and visitors arrived.

"This is XMU 61, clear."

Helios Lookout

Our trip to Helios Lookout was a challenge. The dogs and I were ready at 09:30, but received the news that we would be delayed until noon. We were driven to Whisky-jack airstrip for our appointed time, the helicopter was loaded with gear, and then we all patiently looked up at the clouds that held Helios Lookout in their grasp. A small improvement in visibility sent the helicopter scurrying around 14:00, but it got trapped at the tower as the clouds rolled back in. It didn't manage to come back for the dogs and me until 18:30.

Things got off to a roaring start within minutes of our landing. Jennifer, the resident towerwoman, kept two dogs with her as well. Ted went happily running off with both of them, and Jennifer's dogs seemed intent on showing Ted what a great view there was from the upper deck above the cabin. With one dog on each side of him, Ted was shown how to climb the steel mesh stairs. Only trouble was that he turned too early, walking out onto the slippery aluminum roof of the trailer rather than the wooden deck just a few more inches above. As soon as the first foot slipped, he panicked. Frozen with fear, each foot vainly trying a different direction in which to find foothold, he was a sad sight. Our tower supervisor, Cam, immediately offered to perform the rescue; he climbed over the stairs and slid out onto the roof, grabbed him, and slid him over to where he could be handed back. What an introduction!

By the time everything was unloaded, Ted rescued and Jennifer and her dogs sent on their way, it was very late indeed before a path was formed through the tiny cabin, the dogs were fed and I managed a bit of food for myself, since I hadn't eaten since breakfast.

Even with the abbreviated first day aloft, my first thought as I looked out on the world from a height of over 2,500 metres was how amazing it would be to be an eagle and always look at the world from this perspective.

Living at this altitude had its quirks. A bag of potato chips expanded almost to the point of bursting due to the difference in air density. Drowsiness tapped me on the shoulder just a little bit harder at those low-energy points of the day.

After just two days, an almost full range of weather had visited the mountaintop. It had started out warm and sunny; I even had my basil out on the bannister of the deck until well after lunchtime. But the afternoon cooled off and it got progressively windier. Pattering rain on the aluminum roof woke me overnight, and I awoke to snow the following morning. I found myself looking down at the campers 600 metres below me, wondering what it was like down there and if they had their woollies on like I did.

I finally had all my gear and food stowed away and some much-needed cleaning completed. Jennifer preferred so much furniture packed into this little space that it was difficult to manoeuvre, so I tucked a few pieces into the generator shed. It was easier to go around the wolfhound than to climb over her! This cabin was a trailer as Connaught had been, but it had just two rooms rather than three, this one lacking an office. But it was brighter and cheerier than Connaught, with white walls and light pine cupboards. Several of Jennifer's paintings decorated the walls along with a cross-stitch sampler a previous towerperson must have worked while stationed here, an intriguing bit of history. A collection of fossils occupied the

With a clear blue sky overhead, Helios looks down onto the clouds.

space heater; a little bookshelf held a bit of reading; the kitchen table lit the room with a brightly coloured tablecloth. From just outside the door, a set of stairs extended to the upper deck; inside the cabin, a ladder took you safely into the cupola on windy days.

By now a familiar friend, the Onan generator resided in the gen shed, and the latrine specs were the same as on Whisky-jack: I was back to a bucket. Word had it that the towers "on the bucket" were to get new-fangled propane-burning outhouses soon. Once a month a flip of a switch would turn everything in the holding tank to ash. Cam mentioned that a crew would be coming soon (weather permitting, of course) to pour the concrete base for that facility. Apparently the new outhouse buildings also had a shower stall; not that there would be plumbing or running water, but a place to get out of the wind would be a big bonus up here.

Ah, the views! Helios Lookout is located in the Whisky-jack

Range, about 30 kilometres of conifer-covered hills south of the Brill Range, where I could still spot both Stenhope and Whisky-jack Mountain lookouts with binoculars. There were valleys to my southwest with beautiful winding creeks and meadows, and in the distance I could make out the ranges within Banff National Park. This particular peak (how sad that it didn't have a name) was quite craggy and razorbacked with one level path extending to the east and west, but to the north and south there was quite a drop (as a former towerman found out late one windy night when he took a wrong turn!). Looking almost directly below, I could see the Forestry Trunk Road, Whisky-jack Falls and the adjacent airstrip. A campground to the right hugged the south side of the base of the mountain. I was greeted by fellow mountain inhabitants in the forms of ravens, bushy-tailed woodrats (whose calling cards and odours were many and everywhere) and a mama ptarmigan with teenagers. I spotted some sort of rodent out on the rocks, but without binoculars handy I wasn't able to tell whether it was one of the woodrats or a pika.

The hike from the campground to the cabin was described to me as "four hours for a fit person." After seeing what the hike would actually involve, including a rise of around 600 metres, it did appear possible, albeit a better-than-average cardio workout.

I had the good fortune to be asked by Eaglewatch, a volunteer group based in Calgary that undertakes a biannual count of migratory raptors, if I might be able to help out with the watch, so I decided that would be time better spent than attempting such a hike. Back in Calgary, I had had the privilege of spending a very informative evening with the gentleman who coordinates the count, and he did his best to explain what kind of information they needed. The migration was due to start within the next few weeks, and with a new guidebook at my side, I had a lot to learn before it arrived.

As I had only begun to find out on Whisky-jack, mountaintop weather is like being on a different planet. I witnessed days on end with zero visibility, magnificent undercast sunrises (some days the clouds were below me by almost 300 metres) and watched thunderheads develop in the afternoon sky.

The carpenters and a rapattack crew (trained to rappel from helicopters) arrived later that week to start building the forms for the concrete base that would hold the new propane-burning biff. The crew and their helicopter had been below for quite a while waiting for a hole in the clouds to open and reveal the path. Once they actually got to work, it seemed they had only been here an hour or so when they began to fret that they would all end up spending the night at the lookout. So at the next break in the clouds, the helicopter rushed up and whisked them away, leaving me to pile rocks on the lumber to prevent it from blowing away. Their sojourn would have been longer than that one night, however, because the next two days saw the lookout socked in solid due to low cloud cover.

Even though there were plenty of things to do, it was an eerie feeling wondering if the world was still there beyond that evenly grey view. I couldn't get comfortable letting the dogs out on their own, worried they would forget where the edge was and stumble down the slope. The windows told the story of the cycle occurring outside: the warm air made the snow wet and heavy, then the temperature dropped and the snow stuck to the windows, then it warmed again and the snow slid off, then it turned cold again ...

But Mother Nature would repay us for all of that nothingness with an equal number of picturesque moments. One early-morning glance out the window revealed a pink eastern skyline with the cloud ceiling far below us. I sat up in bed and watched the sunrise first casting pink highlights upon the fluffy finish of the clouds, then turning the mountaintops that same rosy colour as the sun climbed above the horizon. Sun streamed through the windows, and there was barely a breeze. The dogs moved outside to bask on the deck. The clouds were liquid and graceful. When the clouds rose to the west side of the mountain, they flowed downhill through the low points among the peaks. I watched the peaks that held Whisky-jack and Stenhope appear and disappear as they broke through the cloud cover to join us on the sunny side of the world. Patrick over on Whisky-jack Lookout appeared to be perched in the fairytale image of a castle on a mountaintop, wreathed by a cloud. If I could walk on clouds, I could have walked an easy 30 kilometres right to his doorstep and had coffee with him.

Two nights' worth of auroras danced across the skies. Not nearly as vivid as the one that bid us farewell in Jordan, but bright enough to wake me out of a sound sleep. Stretched full length across the northern sky, the aurora was brilliant green, hypnotic in its motion, punctuated with sporadic brilliant yellow and orange.

As I was preparing dinner one evening, the dogs got excited at the sound of an approaching helicopter. We all went outside to see what was happening. After circling a few times, a chopper landed on the helipad. I noticed it was from a company I had seen many times in Jordan. A cheery fellow by the name of Jeremy disembarked and explained that they had been flying around the towers all day, picking the best spots for the new radio repeaters their company had been contracted to install. And yes, Steve, his pilot for the day, sent his regards, remembering me from Connaught. So while Jeremy scoped out my mountaintop, I went over and chatted with Steve for a while. It was nice to see a familiar face.

Upon returning to the cabin, I soon realized I had spent too long chatting. It was 18:20, and I had missed scheds at 18:00. I radioed in right away (it was a bit confusing at first, since they were trying to radio me at the same time) and apologized profusely, feeling bad that I might have scared them.

News from the world below came in audible form as well as visual. Sheep season had opened. The gunfire emanating from the surrounding hills made me melancholy.

The following week's weather was to the benefit of the hunters:

warm winds, sunny skies and great visibility. Meadows that hugged the creeks to the west of us, which had been carpets of bright green grass when I first arrived, had now turned to golds and dark oranges, a tiny little hint of autumn in an area that gets so precious little of that wonderful season.

Southbound raptors began to appear, with more falcons than any other group, but also a bald eagle and a northern harrier. A delightful little flock of ptarmigans, their feathers already changing to their winter white stopped by for a visit. Three horses were a surprise to find at the base of the mountain one warm sunny afternoon; a grey, a buckskin and a bay, all sunning themselves and enjoying a late afternoon nap. I called into the office to ask if people pastured horses in this area and they replied that yes, there were outfitters nearby that kept horses. A day later, I received a call from an elderly fellow looking for Jennifer. He was the owner of the campground and outfitter on the south side of the mountain. He asked if I'd seen any horses about, because he was missing a few including a grey and two buckskins. I had to chuckle when he told me that the horses had escaped two years ago! I had taken out the spotting scope to get a better look and had noted what good condition they were in. I guess if you're going to go AWOL, it's a good thing if you can look after yourself, and this little herd seemed to be doing a fine job.

The carpenters arrived again one fine sunny day and finished pouring the concrete pad. The carpenter kindly called me before he left town to check on the weather and inquire if there was anything he could bring me. I was delighted when he arrived with some fresh milk and a newspaper. The crew arrived just after 08:00, just as I was taking fresh hot muffins out of the oven and had coffee ready on the stove. It was nice to have some company.

After all the changes during the past few weeks, both dogs settled well into the routine in our aerie. The winds would blow steadily from the south or west, sending the dogs to the deck on the lee side of the cabin. I would join them out there with my knitting and watch for raptors. Ddreena contended with sore feet from the rocky terrain even though there wasn't far to walk, but that also meant she was treated twice daily with warm face cloths to soak them, followed by massages with ointment. The early workout at Whisky-jack allowed Ted a smoother transition.

The trick to being successful with chores in these surroundings was to seize the moment while the weather cooperated. And so, on days when warmth and sunshine befriended us, I cleaned and degreased the generator shed, did an oil change, placed a piece of screening over a vent to deter packrats (it was all my fault one got in there, I'd left the door open) and burned toilet paper. With no yard work to do, I truly couldn't think what these mountain towerfolk did to fill time over an entire season.

The warm weather wandered on to new destinations south of us; cold weather and snow soon followed in its place, and outdoor chores were put on hold once again.

Five more fires were added to the season for the district, bringing the total to 94. Two were thought to be hold-overs from a lightning storm more than a week before. The other three were in other parts of the district far enough away that they were reported via different radio channels, so it was difficult to keep up on the news about them. The two that were nearby both occurred in The Gate, which is where the Victoria River passes through the Brill Range. Such is the landscape in these mountains that both fires were relatively close to Patrick but tucked around ridges where he couldn't see them. It was interesting to listen to the radio chatter as they fought the first fire for the next three or four days. My supervisor had mentioned how different fighting fires can be on steep slopes. Not only was the slope and the lack of a nearby water source causing them problems on this particular fire, but the weather kept changing, making it a moment-by-moment decision each afternoon whether to stay and work on the fire a bit longer or get the helicopters to come and get them out before weather would make a getaway impossible.

Before the battle against the fire in The Gate was won, new weather rolled in, accentuated by 100+ k.p.h. winds. By mid-afternoon, a storm front was visible to the northwest. Larry and Patrick made themselves useful and kept the fire crew informed of the storm's progress.

As the wall of grey blocked my view of The Gate, the storm rolled yet closer to us, the winds continuing from the southwest. When the storm reached the east side of the mountain, the warm southwest winds blowing up the south slope met the cold winds from the north slope, and I watched mesmerized as clouds were born. Out of clear air (literally from thin air), mist would form a mere few metres above the ridge on which we sat, swirling about as if testing its new physique, then rising to join the congealed mass above it. Some cloudlets would escape, dashing off on their own agenda, but ultimately they would rise to the commune and be lost to it. And it began to snow.

The next day we had one happy terrier on our hands; if he could have yelled "yee-haw!" he would have, as he cavorted around in the snow, once again scaring the bejeebers out of his human companion as he flitted along the edge of the ridge, defying that 600-metre drop. It truly was a beautiful day deserving of such a dance. All the mountaintops to my west were white, and the pine trees below tipped with frost against the grey and white background. The creek meadows and cutblocks were snow-covered as well, and made watching for smokes the rest of the day a bit more challenging due to the lack of contrast. The traffic along the Forestry Trunk Road below me had greatly increased with all the hunters coming in. I had even watched several small planes land at Whisky-jack airstrip. The road dust rose throughout the day, and I could see headlights late into the night and before sunrise in the morning.

It was as the last glimmers of summer were leaving and these hints of winter were upon us that our biggest adventure presented itself. It was a Friday night, long after the sun had set. The generator was running, and I was just finishing some email when it gave a couple of gasps and quit. With Coleman lantern in hand, I went out to the shed to try and diagnose the problem, but no matter what efforts I made, I couldn't get it going again. After giving Larry a quick radio call, I gave up, took his advice and ended the evening to worry about it the next day when there would be better light. When I re-entered the cabin, I could smell propane, and I noticed that the little meter on my refrigerator was telling me that the pilot light was out. I rushed over to the space heater, opened the side door to expose the pilot light ... and wished there had been one there to see. Damn! I put on my coat and went out to the propane tank, unlocked it from the chains that held it down and gave it a shake: empty.

I had just switched to that tank earlier in the week, thinking it full. So much for that presumption. And they had made mention at evening scheds that they didn't have any helicopters in the district for the next day.

By now it was after 22:00, so I keyed up the radio and gave Patrick a call just to let him know what had happened. If my batteries were run down by morning and I wasn't able to respond at morning scheds, would he report the situation for me? Suddenly grateful for all the blankets and quilts I had brought with me, the dogs and I hunkered down for a cold night. (As an example of the recent temperatures, the bucket in the outhouse had been frozen for a day and a half by this time.) But what a wonderful feeling! Does it get any better than to be snugly wrapped in warm blankets and flannelette sheets, just the tip of one's nose exposed and cold? A quick jump out in the morning, into my long underwear and a couple of extra sweaters, and we were ready to roll. The batteries held overnight, and as I called in my situation at scheds, I was told not to worry; it would be looked after. I concocted a little stove made out of four candles (at below-freezing temperatures, one wick will not boil six ounces of water) and made myself some hot coffee for breakfast, then hot oatmeal for lunch. Just around that time I started hearing radio chatter about a helicopter taking some Fish and Wildlife people up to Snowshoe Creek to see about a bear situation; then the helicopter would be coming back to Shandray camp to pick up a propane tank. Sure enough, around 13:30, I heard the pilot report that he was leaving Shandray, bound for Helios Lookout.

This would be the first time that I'd be required to help out with a tank delivery—I was so excited! The pad where the tanks sat didn't have room for a third tank, so the helicopter ever so gracefully set the new full tank down on the new concrete pad and released the metre-long strap from its belly. I had to remove the strap and lanyard from that tank, take it up to the empty tank and attach it, then wait for the helicopter to come down and hover just inches over my head (it's absolutely amazing how steadily they can do that) while I reattached the strap to its belly (all the while only being 30 centimetres or so away from that 600-metre drop). Off he flew to take the tank down to Whisky-jack airstrip, then he returned to pick up the full tank and set it on the correct pad. After taking the strap off the tank and returning it to the pilot waiting on the helipad and seeing him off with a hearty thank you and a few candies, it was only a matter of reconnecting the hose to the tank, then relighting all the pilot lights. Ahh! Heat again! The only casualty of the adventure was my pot of basil.

I remembered the stories my Dad told my sister and me about being raised in a house with only a wood stove for heat. How his Mother would come in and take the curtains off the windows and add them to the beds as extra blankets on really cold nights; how he and his brothers would race every morning to see who could thaw out the pee-pot first. It really didn't seem like much to put in one cold night, and it certainly brought the trials of my forebear closer to my heart.

So it was from a warm and cozy cabin with the snow piling up outside and the wind whistling by the windows that we celebrated Ddreena's 10th birthday. I don't think too many other geriatric wolfhounds could have made the claim of romping around at over 2,000 metres above sea level at that age.

As the snow accumulated and the days grew shorter, I kept wondering when the season would end. Finally, everyone was thanked for a job well done and told that the towers would be closing over the next week. We would receive a call soon to make final arrangements for each of us. It was such a nice delivery compared to how we got the news in Jordan. Here everyone heard it at the same time, and it came complete with a pat on the back. Even though I kept my ears open, the phone didn't ring the following day, but Cam caught up with me on the next and informed me that they'd be up to get me on October 1, weather permitting. We would be ready, as I had been picking away at closing chores, and there were just the items we used daily yet to pack, with a bit of housekeeping to do as space allowed. (As nothing could be kept outside due to wind and cold, packing actually made it harder to clean.) I began sending emails, letting friends in Calgary know I'd be home soon and letting my family know that I'd be back to civilization for Thanksgiving.

The very next afternoon the phone rang again, and Cam told me that the plans had been changed again. We would all be staying open until the drought code dropped below 300 (Stenhope currently stood at 340; in Lindsay it was still over 400), or October 21, whichever came first. Oh, and they were moving the service days up a bit; if I could have my grocery order ready right away, they'd like to bring it up the next morning. So back I went to the computer and sent off my grocery order, followed by another set of messages to inform friends and family to hold that homeward-bound thought for a while longer.

It was wonderful to hear that I would be staying a bit longer (I still wasn't ready for the season to end), but with such a short deadline for service day, there was no time to ask anyone to send up extra warm clothes and more wool. (As I had officially knit everything I could lay my hands on, knitting withdrawal was at its peak.)

My groceries were ready to transport that next morning; and there they sat. I was fogged in with zero visibility all day. When I radioed the *next* morning to ask whether they were coming, they informed me that they had decided to put delivery off until the following week. This led to a necessarily intense little discussion where I made the case that I wasn't about to accept five-day-old produce and dairy products when I had to make these things last a month, and the helicopter arrived later in the day.

At the same time all of this was happening, I was also in nurse mode. After two days of being fidgety and needy, Ddreena had an abscess burst on her tail. I took her outside to drain it, and by the time I had the job completed, it looked like there'd been a slaughter out on the snow. Luckily, I had antibiotics with me. I never did figure out what caused it. The hole grew during the week until it was a bit larger than a toonie. However, that made it easy to manage, as it was best that the hole stay open and heal from the inside out. What with the cold weather and lack of flies, it was a good time of year to be trying to heal up something like this. Ddreena was so much more comfortable once the wound was opened, that big tail wagged ever so much better, and lucky her, she couldn't see what it looked like. Ted must have felt left out, because he came down with a touch of conjunctivitis in one eye. Ointment came out of the first-aid kit, and we had that under control quickly as well.

Not to be outdone, even though I had had a major dental overhaul before leaving for the towers, I came down with a walloping toothache, and the cold winds outside certainly knew how to make it kick. It was as I was outside one day cleaning and putting fresh bandages on Ddreena's tail, having just put ointment in Ted's eyes, all the while trying to hold a hand over my mouth to stop my tooth from aching, that I began to think it might not be such a bad idea for the season to end.

The snow and cold were sometimes interrupted by warm afternoons, but all in all there was no denying that summer was over. The full moon, now waning, had made it a treat to go out for our "last run" each night. The landscape glowed in the moonlight, whether it was rocky or snow-covered. We were never bored by the weather, because it was a constant cycle of snowy days, socked-in-solid days, beautiful clear days and yet more under-cast mornings that had us standing out on the point and drinking in the rose-coloured spectacle that

The first snows of winter frost the peaks and forests west of Helios.

was ours alone to enjoy. It was just so amazing to look at the clouds so far below ever changing, and to watch Whisky-jack Mountain rise from the mists.

The snow got deeper. If I'd had a toboggan, I could easily have taken a short toboggan ride down to the outhouse; it began to be an adventure to try and find the trail when the winds filled in everything that we shovelled out each day. Ddreena was delighted with the new soft footing and gallivanted like a puppy; Ted had to start bunny-hopping through the drifts. It began to be a greater adventure with each succeeding trip to carry the bucket from the outhouse down the steep slope to the pit, almost up to my knees in snow. Keeping in mind that there was no reason to hurry, it was still challenging to try and feel where I was putting each foot and hoping that it didn't find a loose rock while making sure I knew exactly what that bucket was doing.

I caught myself getting into a routine. I lamented that I was going to have to use a kettle of hot water to thaw out the faucet on the rain barrel to get more water, and then gave my head a shake and went out and scooped up a washpan of snow. But it was getting progressively more difficult to get chores done. The five-litre containers of potable water that were stacked outside the door took over two days to thaw when brought in and placed next to the space heater. It was getting more and more troublesome to walk outside, and I couldn't hang washing outside any longer. While I knew that winter living at ground level could have its challenges, I began to wonder about the logistics and challenges of living on a high mountain ridge for the winter. There definitely were easier lifestyles.

No sooner had all the groceries been put away in the cupboards than the phone call came telling me to pack up. Closing up was an adventure, not only because of the difficulties entailed in stacking all the boxes and moving around in a tiny place, but also because there came a point where I couldn't continue to shut things down until I was absolutely sure I was leaving. I couldn't clean the refrigerator until I shut off the propane, and yet by shutting down the propane I ran the risk of having to fight my way back out to the tanks to turn them back on again, and propane refrigerators are a cranky lot.

The morning that the helicopter was due to come dawned with 40+ k.p.h. winds and driving snow. My few trips to the outhouse that morning had required me to remove the snowdrifts each trip. Even in the few minutes required during my last trip, the drift had reformed enough to prevent me from opening the door to get back out. I hadn't brought the radio with me, and could only think about the helicopter arriving and my supervisor not being able to find me. Those thoughts gave me the motivation to give that door a couple of good hits with my shoulder, and I managed to make a big enough gap to slide through.

The helicopter arrived and dropped Cam and a warehouse staffer off to help me. Because of the high winds, the helicopter couldn't remain on the pad, but went back down to wait at Whisky-jack airstrip. Cam and Beth had come without hats, mitts or scarves, and of course they weren't acclimatized to the thin air. The wind was such that the few precious molecules of oxygen in our vicinity were quickly sucked away. With one person needed at the helipad to sit on the boxes lest they be blown off the mountain, and snow being tracked back into the cabin faster than one could shovel it out, closing was an adventure. The final excitement came when groggy Ddreena, the final addition to the last load, walked right up to the helicopter only to turn on her heels and run back for the cabin. She was quickly tackled and thrown aboard, and we were off.

At ground level? It was a mild and wonderful autumn day.

Cam closed six towers that day. I don't know how he did it.

"This is XMG 640, down for the season."

Luna: A Foothills Retreat

It had been a rough start to our 2003 season. I arrived at the forestry offices in Rumseytown, attended a meeting where I met my new tower neighbours and supervisors, then settled into the bunkhouse until we would be flown into Luna Tower.

Unfortunately, snow arrived before the helicopters could leave Edmonton, and those of us waiting to be taken to our towers were grounded.

Ten days later, I had knitted the first sweater of the season. Although the bunkhouse was cleaner by virtue of the delayed towerpeople in residence, I wished I was getting to know my new site instead. Living out of boxes isn't fun, especially when you take into account Murphy's law stating that the chances you will suddenly need to leave are proportional to the number of boxes you open.

Ted and I were thankful that Ddreena was with us for another season. The trip to the spot where we met the helicopter was ever so much easier on the old girl (I drove my own car to the fire base), and we had the use of a Bell 212 (eight-seat) helicopter, which was much roomier than an A-Star. Rick, previously of Signant and Connaught towers, was now working in the warehouse in Rumseytown. He was such a big help, lifting Ddreena ever so gently from the car

On opening day at Luna, a helicopter brings our gear in for the season.

onto the helicopter. As we quickly slid the door closed, she didn't even have time to protest.

Almost from the moment we stepped off the helicopter onto the site, I was visually overwhelmed. The views both to the north and south were spectacular. The southern horizon was made up of snow-covered mountains, while to the north the hills sloped gently downward into a sea of evergreens as far as the eye could see.

The cabin faced south on an east/west oriented ridge with the site surrounding it extracted from an encircling forest of mature spruce and aspen. It was green and grassy, and it contained a feeling of spaciousness. But there was no time to just wander about with one's jaw dropped open, because there were several journeymen working to check and start the equipment, and I needed to both pay attention to what they were doing and decide what needed explaining.

Whereas in Jordan the only people present during opening were the tower supervisor and the helicopter pilot, in Pineridge they brought the mechanic to connect and test the generator, a gas fitter to inspect and start the fridge, stove and heater, and a communications technician to make sure the radios were working.

It was good to see the Stevenson screen come out. A firepit complete with log benches was snuggled beneath some mature spruce. The garden was adorable with a little white fence along one side, and a cold frame, built from an old storm window and a bit of plywood, was tucked in beside the cabin. The black gooseberry bushes and the raspberries were greening up; irises, day lilies and monkshood were already coming up in the gardens around the cabin.

It was a perfect day to celebrate the beginning of a new season, a bit windy but warm, large cotton-ball clouds sailing through a brilliant blue Alberta sky. Ddreena was already sound asleep out on the lawn, peacefully absorbing the sunshine, while Ted curled up under the desk in the office (his favourite place to den).

As I was trying to get the boxes unpacked, I already had the radio plugged in and my favourite Alberta radio station, CKUA, was adding music to our lives. International shortwave stations would not be needed this season!

The cabin was older than I expected, square rather than the typical rectangular shape of the trailer at Connaught. I jokingly told my supervisor that if the vibrant but bad faux painting job in the bedroom gave me nightmares, I'd be ordering paint. The curtains adorning all the windows were 1960ish pink and purple tie-dye. Maps adorning the walls included a huge six- by eight-foot yellowed and brittle monster that took up an entire wall in the kitchen/ living area. A hand-built willow love seat occupied a corner, the only alternative seating to the unpadded wooden chairs around the kitchen table. On tables, counters, windowsills and shelves, the bedding plants all found sunny homes pending the onset of gardening weather.

Within two hours of my arrival, the first guest walked into the site. I had been told that Luna got around 300 visitors each year (Luna is 42 kilometres by logging road from a campground), but I had hoped for more of a grace period. This particular fellow made me uneasy immediately with his stories of building a cabin "just down the road," the description of his trapline "not far away" and his bragging about how many times he had climbed the tower over the years. That warm fuzzy feeling of solitude began to cool.

Before long there was floor space where there had only been paths around stacks of boxes, some semblance of order in the cupboards and office, and there was even water in the rain barrel. The cabin was beginning to look like a home.

The cupola had been replaced the day before I arrived. What a mess it was! One window had a starburst fracture. Blocking the view in its direction were strips of masking tape used to stabilize it temporarily against the wind. Despite having to clean the mud out of the drawers in the firefinder table (there was mud everywhere inside—where had it been stored?) and the walls, it took me only an afternoon to put it in almost working order. The firefinder was now oriented and levelled (Stenhope Lookout could now be found in the correct location at 243°, not the approximately 130° it had first shown), and there was an almost-functional map on the wall. I could now see out all windows. The base station radio that the communications tech couldn't find eventually emerged from under the garbage left behind by the installers. I reported the find to my supervisor, who asked if I could wire it up myself. I had to laugh because all the wiring leading up to the cupola was tied in a wild dreadlock-like knot under the floor. A hole hadn't even been drilled to pass the wires into the cupola yet!

Until they returned, I had one hand-held radio that had to follow me everywhere (except to the outhouse—no way). That said, the new cupola was an improved model with updated features, such as better window frames. Once all the deficiencies were dealt with, it was going to be grand.

Although the cabin was old, it seemed to have been well looked after. The previous tenants had left behind a bit of everything: food, vegetable seeds, candles. It was fun finding this and that tucked away in cupboards and drawers. It also wasn't long before I had some forest denizens knocking at the windows. The spruce tree that stood just outside the east window allowed both a red squirrel and a whisky-jack a place to sit while they waited patiently for me to pass seeds out to them. A pile of broken sunflower seed shells right below the window told the previous year's story. I must admit to testing the manners of the cheeky squirrel by offering a few bits of trail mix right out of my fingers, but the whisky-jack seemed a bit more leery of this approach. Mrs. Whisky-jack was already accompanied by two young nearly as big as she! The young birds were quite comical as they investigated the world around them. They picked at shiny objects, threw sticks around and did their best to convince their mother to supply them with a bit more to eat rather than having to find it on their own.

While all this settling in was occurring, the summer-like weather that greeted Ted, Ddreena and me upon arrival passed, losing its way to snow over the next three days. The temperature only reached single digits, and some days were just plain nasty due to the wind. While I had hoped to start enjoying the night skies with the impending lunar eclipse, that night found the sky a deep navy-grey with huge snowflakes coming down, frosting all of the trees in a distinctly Christmas-like way. Can there be anything more peaceful than a coniferous forest draped in new snow? So quiet and indigo blue. The two inches of accumulation was almost melted by noon the following day.

The dogs and I delighted in the greater opportunity for walking here than we'd had at Connaught. We took a first jaunt down the road while snow still covered the mud, and evenings found us following the trail through the woods westward to a neighbouring cutblock where there was a grand assortment of ungulate scat, which I presumed to be that of moose and both mule and whitetail deer.

Three adults and four children made up a group of visitors on our second day. Two days after that, 33 more arrived. This was certainly not Connaught. I began to wonder if my job title needed to be changed to "Towerperson and Tour Guide." I thought of Larry up on Stenhope and how much he enjoyed his visitors. I could only think that the three whitetail does who travelled through late one afternoon were not only much quieter, but more welcome as well.

Upon further examination of our ridge during walks, I decided this forest reminded me more of mid-Ontario forests than our muskeg home of last year, perhaps like those between Barrie and Algonquin, allowing that Alberta has consistently fewer species of most everything. The surrounding area was populated in aspen and spruce; to find pines I had to look hard. Although we passed tamarack on the highway to the airbase around 25 kilometres away, I never did find any in the immediate vicinity. Cow parsnips were abundant and just starting to break ground, with grasses and strawberries in equal pursuit. It was a wonderful surprise to discover a patch of rhubarb, and several clumps of chives coming up in the garden. A woodpecker (by its call perhaps a pileated) told us of his patrols in our woods, although it remained hidden from view. During snowy times the chickadees would fill the forest with their chatter. A robin graced our lawn, ravens cruised high over the hills. My identification skills proved weak as a dull olive sort of vireo/warbler challenged me, and when I spied my first raptor, I could only go with my initial thoughts that it was an immature female northern harrier.

The icing on the cake was when we came upon our first yellow-rumped warbler. How wonderful it was to hear his familiar tune! It made me smile and brought a tear to my eye. I had been wondering these past days if I would ever find myself feeling as settled as I did at Connaught, and felt that this was the sign I had been looking for. I quietly imagined he might be on his way to Sandal Hill.

As I wasn't spending much time in the cupola yet, I had a bit of fun decorating the outhouse. I had brought a big stack of postcards that friends had been sending me, trying to entice me to visit the U.K. and bringing me a variety of views in case I got "tired of looking at the trees." I wallpapered the back wall with all the postcards and made a little shelf into a library, thus finding an appropriate spot for all those old *Reader's Digests* I had brought with me.

Back in the cabin, it was a battle trying to get a steady signal on my bag phone to achieve a stable Internet connection. It was baffling why it had been so easy at Connaught, while here, where there were many more cell towers and communities, I couldn't get a dependable connection. When the comms tech returned to install the base station radio, he generously lent me a yagi antenna (a tubular aluminum type). Upon finding a steel pole lying behind the gen shed, I propped it up against the cabin and tied it off, attaching the yagi as high as I could. Although it meant I had a telephone line running across the cabin, I finally achieved a stable connection. Success, and another chore off the list.

"This is XMG 722, open for the season."

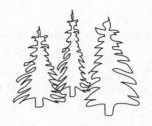

From Popsicle to French Fry

How could Alberta's weather be so different from Ontario's? When I first arrived in Alberta, I thought that resident Albertans were exaggerating their weather claims. Luna, like Calgary, so closely hugs the Rocky Mountains that it tends to experience the phenomenon I refer to as a micro-climate. Its "micro-ness," I feel, is derived from two different realities: one, what is happening over your head at one particular moment isn't necessarily occurring a kilometre away; and two, what is happening at this particular moment isn't necessarily what is going to be happening an hour from now.

With those observations forefront in my mind, I witnessed on my arrival at Luna a confirmation: delays due to snow, followed by cold, wet, windy days, culminating with a bask in 27°C temperatures, all one right after the other! As the last hurrah of winter gave way to the first breaths of summer (did you notice spring was missing?), the Duty Officer duly declared Luna, along with a few other towers, to be on high hazard level. And so I went from being earthbound to aloft in my cupola once again. Ddreena proved her memory remained in fine form—she was soon howling up at me in the tower on the same schedule as she had the previous summer.

During that last week of May, the last of the mountain look-outs in the district opened: Helios, Vantage Peak and Whiteley.

The complete set of 11 was now operational. The towerpeople in this district were vastly different from the group up in Jordan. Up north, it was almost impossible to get the other observers to stop phoning and talking the evenings away; here it was almost impossible to get any of them to speak at all. Being the "new kid on the block," I longed for insight into how things were done here, but no one was willing to share.

With so little time on the ground, it seemed agonizingly slow to get the yard in order and the gardens planted. But it wasn't long before the snow peas that I had started back in Calgary began to take hold, and the cherry tomato looked as if it might succeed. I patiently awaited the first spinach and lettuce seedlings to break ground; the Swiss chard would be planted soon. Johnny-jump-ups (violas), African daisies (osteospermums) and portulaca, the few flowers I brought along as a treat to myself, were all assigned their spots in the gardens.

At this time of year the forest changes daily, with the trees waiting for just the right moment for their buds to explode. The aspens went from closed bud to open leaf within one week as the groves around and below me donned that golden halo-like glow of catkins extending, then on to the brilliant lime green of the new leaves. The spruce buds lagged behind, but many under-storey plants were now poking through last autumn's blanket of leaves on the forest floor. The species list for this site was beginning to look challenging.

The birds continued to delight both my eyes and ears—the nut-hatches with their gentle nasal declarations and my personal favou-rites, the chickadees, singing their spring song. I spied a pair of yellow-bellied sapsuckers (which I had up until now thought were hairy woodpeckers) who seemed to call this ridge their home. I remained convinced that the woodpecker I could hear was a pile-ated. A rose-breasted grosbeak graced us with his presence for a day; the population of yellow-rumped warblers continued to grow; the juncos spent time finding the exactly right pieces of cured grass to line their nests, and Mama Whisky-jack continued to teach her two children how to forage. When a mature female northern harrier cruised the forests around the tower, my heart soared. Just as it was when I heard the warblers' songs, the harriers' presence brought me back to Connaught. As for other types of winged inhabitants, the lovely lilac butterflies I became familiar with last year were also here (probably a different member of the same family, but undoubtedly a blue of some sort), along with a few new species.

Ted created a job for himself, becoming our little doorbell. Each time a quad would pull up outside of our gate, he would begin to chirp, allowing me a few moments before the visitors rounded the bend in the path and came into view. So I was just a bit surprised when I heard him bark and looked out of the cupola window to find him facing into the woods instead of down the lane! I couldn't find what it was that he seemed to hear or see, which was just a bit disconcerting. Later that afternoon I saw something move in the clearing just below the helipad. At first I thought it was a bobcat or lynx, with only the motion of his tan coat visible against the still brown grasses. But once I got my binoculars out, it turned out to be a whitetail doe, lying peacefully on her side enjoying a midday siesta. After an hour or so, she rose and gradually grazed her way out of view. Later three does travelled off to the west as the sun began to sink. After several repeat performances of this trek, I surmised that this was their daily sojourn, from the west, up over the ridge, and off to the north for the day, and then back over the ridge and off to the west for the night. Although they weren't sure what to make of Ted in the beginning, they eventually accepted that he would only go to the edge of the site and that there was nothing to fear.

Luna Tower had received a bit of rain again, but the top of the hill didn't seem to get the same quantities that fell around us. Friday was flatly overcast, cold (a low of 0°C) and rainy throughout the day; Saturday was a constant parade of towering cumulus clouds. I watched as curtains of rain challenged me by blotting out bits of my view. Twice I looked up to the splendour of rainbows, accompanied by sparrows giving their "storm's over" serenades. At the day's end, as I was having thoughts of heading to bed, I took a final glance out the window to find the most magnificent sunset; the dark greys and dull blues of an overcast sky, dimpled and uneven, cracking on the western horizon where the setting sun was peeking through. Those rays caught a few small, fluffy cottonball clouds that were floating independently below the ceiling and turned them fluorescent pink; they shifted moment by moment, catching bits and pieces of the cotton-candy ceiling, turning those pieces that same magnificent colour, as if someone had taken a dull blue cotton candy and given it a final frosting of pink. In a few minutes it was all gone as the sun sank below the horizon and the dull colours of dusk again predominated.

It wasn't all that far into the season, near the end of May, around 16:00, when suddenly a huge black smoke appeared on the northern horizon. I quickly called Larry at Stenhope, who had years of experience looking over this area, to make sure that in my newness at Luna I hadn't just seen a flarestack, although I couldn't find one in my record book that matched this particular bearing. Larry said that he was fairly sure it was a flarestack (as they can be prone to suddenly come up like that and disappear just as quickly). Jennifer from Helios Lookout joined the conversation, telling us that Bill at Brill had called in a smoke just 15 minutes previous (Brill was on a different radio channel that I hadn't been scanning and his tower was much closer to the location of the smoke). In just those few minutes, the smoke had grown noticeably larger.

As I switched my radio to scan all channels, I heard that Bird Dog 30 and his team of tankers had been dispatched from Kinsey with an ETA to the fire of 15 minutes. The Ducks, led by Bird Dog 3 and stationed in Long Lake, were just taking off; there were now three helicopters in the air, complete with rapattack and HAC

(Helitack) crews. An incident commander had been named to co-
ordinate everything; water sources and helicopter landing sites were
identified; fuel cache locations passed; and equipment transported
to the site near the town of Jackston. Amid the chatter there were
brief mentions of evacuations, dozer lines and keeping the fire south
of the river.

Pineridge Wildfire #49 was born.

In what seemed to be less than an hour from when I first
spotted the smoke, the fire had doubled in size from 45 hectares to an
estimated 90. The column of smoke extended all the way to the
clouds, stretching out under them as if there were a horizontal
plate of glass separating them and preventing them from mixing.
The following hour saw the smoke grow visibly smaller until it was
officially declared "being held" at dusk. The next morning there was
no trace of smoke left on the horizon.

The sudden heat subsided at that time and much-needed showers
followed. As the rain barrel filled, the hazard level dropped and
allowed me time on the ground, a real treat because I'd been on
high or extreme hazard level more often than not. Once again it
had been so long since I had addressed life on the ground that I had
to wander around the yard a bit to finally remember what chores
needed to be completed.

Once organized, I finished planting the last of the gardens (not
that gardening is ever finished) and, remarkably, each day that I
took hoe in hand, a flash hailstorm ended the day.

A shrub showed off its little white and pale pink flowers, and
several wildflowers visibly struggled to burst fattened buds; these
first spring flowers heralded the beginning of June. Pollen blow
began: the male conifer cones shed such copious amounts of pollen that my visibility in the tower was reduced to 15–20 kilometres from what should have been 40. As the breezes came through
each day, the yellow clouds would blow up from the ridges. Since

While it appears that the smoke from the fire is spreading beneath the cumulus cloud, the smoke is actually creating the cloud.

it was warm enough to have the windows in the cabin open, most objects inside had a yellow dusty glow to them, and the water in the barrel was covered in yellow frothing muck that smelled as though it were fermenting.

The first hazy days also rolled in, and I was not certain whether they were caused by smoke or something else. Late in the day it would get quite pretty, with the individual ridges of mountains and foothills turning their unique shades of grey, the scene becoming almost one-dimensional, with wavy bands of varying greys painted across a wide horizontal canvas.

Another substantial quantity of rain came over the next few days, the longest rainstorm since opening, and the level in the rain barrel rose again. Cleaning water (not the potable supply) had almost hit rationing stage prior to this, with contingency plans in effect to do dishes every other day and leave laundry for a while yet; if I had to forego bathing, then I prayed for no visitors. The storms upped the barrel to half-full, enough to at least get us through another couple of weeks, especially since the gardens wouldn't need watering in the near future.

Only two people came that particular week, the first fellow arriving in the early evening catching me totally unaware, since most visitors arrived midday. I was still having difficulties getting used to having people arrive unannounced. This fellow was dressed in camouflage, and he asked me if I had seen any signs of bears. He then went on about how he was one of those totally "moral" hunters who would never shoot a sow, only boars, rah, rah, rah ... Did he really think I would tell him if I'd seen one? And then he went on to claim that he knew just about every bear in the area—so why did he need information from me?

The second fellow was also a hunter, bragging that he'd gotten his bear. "Too bad," I replied, and walked away.

The entire season would end up being a lesson in the acceptance of people whose perceptions of what our responsibilities to the forests entailed were vastly different from mine. Hunters and trappers formed a sizeable proportion of the visitors to Luna, and I had to learn to "put on the mask" to get through conversations with these people. More than anything, up to this point in my life I had lived a relatively firearms-free existence and this philosophy was important to me. One brief experience with a now ex-spouse who had insisted on breaking all laws for storage of his firearms had left me even less tolerant. It wasn't just my aversion to having armed strangers walk into the yard, it was the attitude of far too many of them that troubled me. There was never a sense of reverence, of being thankful for a supply of fresh meat. There were the types whose eyes would glaze over as they described how bears scream as they die, and freckle-faced 14-year-old girls with perky ponytails anxious to "blow away a deer." Equating the impact of taking bears from the ecosystem with, "Oh well, they're just pests anyways," they punctuated their lax statements with casual shrugs of their shoulders. There were others who would already have their elk or mule deer to satisfy their tags but stay on to shoot a few whitetails ... just for the fun of it. Was everyone of this mindset? Of course not, but their numbers were disheartening.

But even more difficult to rein in were my hopes for solitude over the summer. For me, a large part of the satisfaction of work at a tower is the opportunity to be alone; to think, to watch, to enjoy a sense of intimacy with nature. Having people walk into the site so relentlessly for the entire season disappointed me. I would take a rare mid-afternoon nap on a low hazard day, only to awaken to find a group of people standing in the doorway of the tent staring at me. I'd be deep into my writing in the cabin with the noisy generator running, only to have someone startle me with a knock on the door. I had to remind people that this was not their space to do with what they liked—it was all so very different from what I had hoped and imagined for my time at Luna. As I tried to understand why visitors evoked these emotions in me, I happened upon this quote that I lauded for its clarity:

> *Some of the tourists ... come looking for peace and beauty, but they must also bring such desirables with them or they will never find them here. The words of those who do gleam like sunlight on flowers. Gradually a character unfolds, opening up as do poppy petals in the morning light. We are all tourists really, discovering new regions in ourselves and in others.*

—Gilean Douglas

I remember with a certain droll resignation watching from the cupola as a fellow brought his dog on-site and allowed it to urinate on the raspberries. His thoughtlessness saddened me, because treats such as fresh fruit aren't something a towerperson at a fly-in tower can just run out to the store and purchase. It wasn't many days later when a woman walked in and decided she should enjoy those same raspberries, and proceeded to pick and eat them. I fumed at her insolence—then remembered the dog incident. There seemed a certain pathetic justice to that sequence of events.

Not that there weren't also thoughtful people. One family that visited several times over the summer brought me fresh cherries and grapes—food of the gods when everything in my refrigerator was over three weeks old. But such people were few and far between. Most came with the "zoo mentality"—to stare at the strange woman at the fire tower. They needn't have wasted their time and gasoline.

A surprise visit from a helicopter brightened one lovely afternoon, and I was delighted to find Thomas, an acquaintance from my time at the school in Hinton, was on board. It was wonderful to see him again and share how my new assignment was going. He has always been and continues to be so kind, keeping in touch and being a contact when I'm having trouble with this or that. He fixed a few things that were still amiss from the recent cupola installation, while his crew cut down a few lovely mature trees that were apparently too close to the Stevenson screen. A gentle reminder of the human propensity to miss one familiar tree while standing in a forest containing thousands.

"The Girls," as I dubbed my three whitetail does, continued to travel past the site. I began to wonder if I might see Mom with a new baby, since the twin yearling does were now on their own. The raptor I had been wrestling to identify returned. Whatever the species, it was a talented glider and delighted in spiralling for hours in the updrafts north of my ridge. It seemed to be the right time of the year for soaring, as the swallows arrived as well.

As if it wasn't enough for humans to have bad days, the squirrel that continued to believe that it could train me to feed it arrived, all buggy-eyed and anxious, repeatedly throwing herself at the window. Although I did relent and offer three raisins over the morning (I was trying to break it of this habit), when I opened the window she lunged at me, ending up in the cabin. She then had to make several attempts to find the open window to get back out (wonderful, I thought in that split second, that's all I need, a loose squirrel in the cabin!). Once back outside, she peered through the windows, following me as I moved about the cabin. When I was at the kitchen sink, she would be kited on that window screen, when I walked to the refrigerator, she would toss herself at that window, all the while her poor little chin chattering away as counterpoint to the panicked look on her face.

After my experience with the biting bugs at Connaught, I thought it would be clever to bring along my screened tent to provide a small bug-free space in the yard. To celebrate the peas, spinach and lettuces all breaking ground, I put up the tent, pulled out the chaise lounge and sat—radio, dogs, iced tea and mosquito coil at my side—and read for a while. I so enjoyed my afternoon that when they announced at evening "scheds" that we were at low hazard level again the next day, I thought perhaps I'd try repeating that siesta. For as much as people think that towerpeople have large amounts of time on their hands, relaxation time is actually pretty hard to come by. I try my best to enjoy it to the fullest when it does miraculously appear.

The low hazard level stretched to four days in a row while I accumulated 17 millimetres of rain. The rain barrel was now three-quarters full once again.

I spent another day enjoying the screened tent, then the next day brought off-and-on showers and hailstorms. Mother Nature followed that up with a perfect butterfly day: clear skies, a gentle breeze, the aroma of warm spruce trees and as many butterflies as the lawn could hold. There were swallowtails, white admirals, blues, tiny little butter-coloured ones and two different varieties clad in orange and brown. The dogs and I walked down to the cut-block to find that the wild purple clematis bloomed throughout the woods, like graceful mauve fairies hidden in the shade of the mature forest. The baneberry was now in bloom, as was the false Solomon's seal and sunny yellow arnica; wild strawberries adorned the lawn near the helipad, and the rich blue lungwort would soon be joining in. The garden allowed me my first batch of stewed rhubarb. By the week's end we were back up to high hazard level, leaving me to watch the butterflies sail gracefully though the yard from my perch.

I called my first smoke around this time, but it ended up being a temporary flare. An oil company had warned forestry about it, but the only towerperson who was informed was northwest of me. Even though I was a bit angry with myself for calling a flarestack, much good came of it. First, it wasn't actually a fire; second, it got me past the stage fright that comes with the first detection message of the season, when my adrenaline runs high and I fret over being a babbling idiot on the radio. In addition, I got all the details down in good time, and my bearing was dead on. My estimate of distance was off by five kilometres (I said 25; it was actually 30), but I still didn't have a proper map or a good feel for the area, so I didn't consider that too bad.

Another bit of excitement came our way when a group of tankers left Rumseytown on a return trip to their base in Kinsey. The radio keyed up, and Fred at Whiteley Lookout informed me that I should be able to see one of them at any moment returning to Rumseytown with a smoking engine. I turned my binoculars to the west, and within a few moments spotted the plane as it dumped its load of retardant. That in itself was interesting, as I had never seen that done before. The bright red retardant left quite the veil of colour across the sky. The plane passed my tower closely enough for me to clearly see that the right propeller wasn't turning, although the pilot had successfully quelled the smoke. I watched until it disappeared from sight. A call into the radio room later that day confirmed that it had arrived safely.

I was still adjusting to the early onset of darkness in the evening (although at this time, being early June, there was still a bit of light well past 23:00), so I would make it a point to sit on the front step just before bedtime and observe the sky. I hoped if I made the effort to deliberately absorb the darkness, I would grow accustomed to it. The recent clear skies had offered us rich vivid deep blues, and the waxing crescent moon shone brightly, cresting the trees well before bedtime. Mars was easy to pick out because it was travelling in close proximity to earth; it was easy to identify with its alternating red and white glow. We were often joined by a bat, a new member to our circle of beings on the ridge, that circled the cabin at dusk in search of insects.

I decided it was time to give Connaught a call and had a great chat with the young woman who was the new observer there. She seemed a delightful young woman, and being a resident of the area, she wasn't surprised by what she found on the muskeg. She reported that the caribou did not stop by and that she was down to just one bunny, but she had spotted a great horned owl. She wholeheartedly concurred that there was something truly magical about Connaught. It was so good to hear that it was in the hands of someone who appreciated that.

I overheard Helios Lookout call Whisky-jack Mountain Lookout to say there were white pelicans flying over the Whisky-jack River. I had lived in Iowa briefly right after graduation, and had watched the flocks of pelicans resting on the small ponds along the highway. To manage such a long journey twice a year must be exhausting.

Outside the office window the tall aspens were dancing a wild hula, waving and bending their bodies madly in the strong winds. Picture a beautiful mid-June day, warm, with huge cumulus-cotton puff clouds sailing across the sky, some depositing a shower along their travels. One particular cloud, backlit by the sun against a light green landscape, had its veil of showers glow a vibrant pea-soup green. Later yet another shower reminded me of a laser light show: the showers angled to the right, with the sun shining through the veils from the left forming a crosshatching, all combining to create a grey, white and pale blue tartan between cloud and land.

Another shower whooshed by our ridge, bidding us farewell with a rainbow that seemed to end just a short walk down the road. I wondered if it might be a sign to calm my doubts about whether I was meant to be at Luna this summer. Was my pot of gold on this ridge? Having settled in so well at Connaught, I thought that other sites would be equally as welcoming, but Luna stood in dark contrast. A feeling of rejuvenation and belonging was missing here.

Our days of watching trees dance and clouds gallop turned into a struggle with cabin fever when cool temperatures, rain and zero visibility set in for three straight days. Other than to collect weather data, there wasn't much of a reason to even go out the door. On the third day, when the skies finally cleared in the late afternoon, the dogs and I literally dashed out the door to greet the world again, making our way down the laneway to see what had changed. It wasn't until we returned and I could hear the radio squealing that I realized we had missed evening scheds yet again. It was worth it.

That week held so many special moments. The dogs and I startled several whitetail that happened to be crossing the road as we rounded a corner, and we found a little trail that turned into a deer track that brought us to a moist glade populated in ferns. At first I thought we might be too late, but with some patience and searching, I collected the most delectable handful of fiddleheads for dinner. I can't begin to convey how scrumptious they were—wild, fresh, steamed to perfection and dressed in a few drops of cider vinegar and butter.

Another day I opened the door to find myself staring at two bull elk looking very elegant in their velvet, grazing at the helipad. I quickly grabbed my camera, but only managed one pathetic shot before they disappeared into the woods, shy beasts that they were.

Not five minutes after the elk left, a helicopter made a surprise stop with a tank of propane. The HAC crew on board had a botany buff among them, and he and I enjoyed walking around sharing thoughts on what was what around the site.

Bohemian waxwings arrived on the ridge, sailing past the cupola windows where I was working on my embroidery in the afternoons. I also pondered over some sparrow-like birds that were dining high in the aspens, marked by speckled breasts and prominent wing bars, but I just couldn't seem to get a firm ID on them (I later identified them as female white-winged crossbills). Chipping sparrows were everywhere; I decided that a warbler I had seen on the road a while back was of the Tennessee variety.

I delighted in the success of the garden, except for the rebellious tomato plant. Somehow, just as a cluster of flowers looked like it might soon open, it would get broken off. While talking with the family who manned Luna the previous year, I learned they had a visitor they dubbed Martin the marten who used to help himself to the Swiss chard in the evenings. I wondered if it might be him.

The snow peas were now blooming, the sweet peas were uncurling their first leaves. The young plants I thought might be delphiniums were actually wild white geraniums, and the Asiatic lilies that someone had planted here had grown tall and were beginning to extend their buds. I was curious what colour they would be.

As for the wild garden Mother Nature blessed us with, a wild pea or vetch was growing in leaps and bounds, the buds turning pink at their tips; the baneberry, false Solomon's seal and arnica flowers smiled from their shady spots. A little patch of western Canada violets just beside the raspberry canes had begun to bloom; they were so much bigger than those at Connaught! And, oh, so many butterflies floated everywhere.

The mounds of two anthills of the huge Alberta variety inhabited the area around the helipad. Nothing like what exists in Ontario, the largest of these "cities" was easily one-and-a-half metres in diameter. Now with all the warmth, there was always much activity as thousands of the tiny city dwellers rushed around.

With the rain barrel replenished (it conveniently overflowed and carried off the rest of the conifer pollen scum) and a breezy blue sky, the clotheslines filled with clothes and bedding. There is just nothing like going out to a clothesline and enjoying the fragrance of clean laundry.

For so many of the regular chores, the big difference between how things are done at a tower versus the city comes down to a lack of two things: electricity and running water. Although I laughed at myself during the first couple of days when I found myself reaching over the sink in search of the taps out of habit, I truly didn't miss either. In fact, it's fun to try to figure out the best alternative ways to do things (anything requiring hot water does need a bit of planning since large volumes take a while to heat up on the stove).

I found that with one season's experience under my belt, my priorities fell to managing the water in the rain barrel and to disposing of the volume of garbage I generated. With these goals identified, it was then just a matter of deciding the best way to achieve them.

I set water usage according to how much I had, relinquishing control to Mother Nature, because only she knows when it might next rain. When there was lots of water, I scrubbed big things like sheets and dog blankets on the washboard in the big galvanized tub; when levels were low I only washed shirts and underthings out in the kitchen sink.

One thing that made a big difference in the volume of water necessary was limiting the amount of soap (I can proudly say that where the grey water settled outside the cabin the grass remained green). Laundry can get clean and only need one rinse; dishes don't need to be rinsed at all (and they sparkle—really!) and only need to be done once a day. I can shower in about one gallon of water. Potable water is delivered in 5-gallon (22-litre) containers, which I used at the rate of four to five a month.

I delight in producing as little garbage as I can, a predilection from the skills I honed kayaking with zero-impact campers. I choose foods that have little or no scrap or peelings (rice is better than potatoes, for example), including the packaging. I burn scrap paper, thereby accumulating only a 60- by 60-centimetre cube of garbage, with glass and cans separate for recycling. Items that I can't clean to an odour-free state, like meat trays lined with absorbent pads, go back into the freezer until service day so that they don't rot and smell.

In Celebration of Long Days

If I had been back in Calgary, I might have celebrated the solstice among the thousands of people drumming at the Olympic Plaza or enjoying the Stoney Tribe's summer solstice celebration on Nose Hill. In the foothills of the Rockies where Luna Tower sits, I was the lone human among so very many plant and animal companions. It was a cold, windy, rainy day on the ridge with a meagre high of 5°C. When passing our morning weather, Jennifer on Helios Lookout greeted us with a hearty, "Ho, ho, ho," so we knew what adorned her mountaintop home when she awoke that morning. With the shorter days of this more southern latitude, there just didn't seem to be as much to celebrate, because the sun would not set at midnight as it had at Connaught.

The days leading up to solstice had been a mixed bag, one day baking us with a high of 28°C, followed by two days that brought us almost 35 millimetres of rain. In spite of the rain, there were fun moments. The Girls arrived just as I was wrestling Ted through a bath. The more curious one peeked out from behind a spruce tree just on the far side of the tower, blowing and stamping her foot to see what kind of a reaction she could get. Or possibly Ted had a pact with them to divert my attention long enough for him to escape. Neither would admit to any such plan.

One low hazard morning I decided to remain in my pajamas until late in the morning, looking the total scruff with my knee-length nightgown hanging from under my fleece shirt, my pants pulled on over bright fleece socks, hair askew. (As towerfolk never get days off, little changes can help avoid the monotony of perpetual routine.) So, of course, two men came walking in, both nice enough fellows, and I apologized for my appearance. They seemed not to care, and we all sat on the steps and chatted for a short while.

At Connaught, I had tried to stay connected to society by not allowing my volunteering to fall by the wayside. I busied myself knitting sweaters for the local hospital auxiliary. Now I hoped that Yukon Learn would find me another student to tutor, but it didn't happen. I applied with Rumseytown, which had an Adult Literacy group, but it turned out they didn't run a summer program.

Visitors began to show up more frequently as the weather improved. The third group to arrive one particular day drove an SUV, the first car-like vehicle to make it to my gate. It sent chills through me that there was yet another vehicular method by which to travel to Luna. How many more people would be coming?

The manner in which cupola occupancy was directed in Pineridge Forest (and in other districts as well) was changing. In Jordan it used to be directed strictly by the individual tower's FFMC (fine fuel moisture code); here in Pineridge it was set each day by the Duty Officer. How each DO dealt with it varied, and it took a bit of adjustment. I found it more natural to flow with the weather patterns, moving up through the five hazard levels more smoothly than by the office-dictated method, which at times leaped back and forth between low and high or extreme.

A helicopter touched down one day loaded with tree-planters who were to replant a cutblock down the road from me. The surface of the cutblock was too rough for the helicopter to land, so they borrowed my helipad, and quads were dispatched to Luna to take them from my yard to the cutblock. It was one of those rainy/sunny/rainy cold days, and they were a sad, muddy, chilled bunch of Gortex-clad young adults who trudged back into the yard at the end of the day. Several gratefully accepted some hot cider drink before being whisked away. The helicopter arrived early before they got back from the site, so the pilot sat and chatted with me for a while before taking off to see what his charges were up to. I quite enjoyed the conversation. These fellows know what I do and what happens at the towers, so it's possible to have an engaging chat, unlike the conversations I endured with the quadders who repeatedly inquired if I was lonely, along with other thoughtless questions while staring at me as if I were a sasquatch.

At Luna, near the end of June, the forest became thick with mottled-brown little moths, and from the tower I could see clouds of them as they hovered around the treetops. I was startled on a first-of-the-morning trip to the outhouse when I opened the door only to meet hundreds of them trying to exit simultaneously—that woke me up! For the next couple of weeks, I would stand back as I threw the door open allowing them to fly away without a collision.

Other temporary visitors were giant solitary hornets. One of these fellows would suddenly fly through an open window in the cupola and work himself into a dither trying to get out. They were usually directionally confused or stunned for some unapparent reason (it was warm enough and they didn't crash that hard). I thought it best that I neither sit on them nor inadvertently stitch them into my embroidery. Luckily, it was easy to convince them to climb aboard broom bristles, and then out the window they'd go. Whatever their mission, they needed to get back to it.

Igor joined our family for the season. Igor was what I came to call "our railing spider." (She was actually an Orb Weaver, *Araneus* sp.) Tucked under the north railing of the porch, she appeared with the warm weather and disappeared with the first snow. She was always an affable gal, minding her own business, and never once invited herself onto my lap or neck (unlike her friends in the cupola). At the beginning of the season she would weave a new web every two to three days, and in return for this fine new piece of art, I would occasionally supply dinner of a housefly or horsefly that had become particularly annoying. The speed with which she would rush to her dinner was incredible. Within seconds the prey would be mummified and then carried up to a secret corner that served as the pantry. While quietly waiting for new prey in her funnel-shaped "garage," she always had one leg placed strategically on a supporting strand. When delivering a meal, I would give the web a couple of taps and then fix the dinner offering on the web.

As the summer wore on, she would seem to hibernate after particularly large meals only to emerge from her "garage" at least two dress sizes larger and progressively in less of a hurry to get a new web spun.

I couldn't believe my eyes the morning I came out of the cabin and found the remains of a swallowtail butterfly hanging in her web. Long dead and its wings in shreds, it looked pathetic hanging there, its huge cigar-shaped yellow and black striped body dwarfing Igor's rotund physique. That old human trait of qualifying what should and should not be quarry came out in anger, which I vented directly at Igor (who was I to decide on what was a fit dinner?). Not a single fly did she get from me for the rest of the season. Not that she cared, as she hibernated for more than a week after that feast.

Crab spiders were abundant at Luna as well, tucking themselves into garden flowers or aspen leaves. There they would sit for hours on end with their arms spread wide, waiting for the first nectar-eating prey to stick its head in. Equally as fun were the vivid black-and-white jumping spiders—I'd be surprised if everyone hasn't spent some time contemplating these furry little fellows in childhood. They aren't web-spinners, but wait for prey to land near them, at which time they leap upon it. They tended to warm spots like the tower structure or the metal water tank while waiting patiently for their next meal.

The various spiders who claimed Luna's cupola as their home had a warped sense of humour that was never evident in Igor. The webs I had to dismantle always seemed to be between the spotting scope and the surface of the firefinder. Seldom larger than a dime, these arachnids would hide behind the bolts at the seams of the cupola, then scramble up until they were directly above my head. From this launch point, they would drop down onto my neck or onto my hair where they would start a second descent from my bangs onto my embroidery. I swear I could hear them giggling as I tossed them out the cupola window. Thank goodness I've never been an arachnophobe.

On one of our daily excursions along the road, the dogs and I came upon the proverbial "when a tree falls in the woods." A gigantic old spruce had succumbed to a weakened trunk and likely a windy day. The scene was fascinating. The 15 or so metres of the trunk stretched along the forest floor; amputated limbs from neighbouring trees joined it there in repose. With all the logging and young-growth forests that we so often walk through today, how seldom we get to see a tree that has just died of old age. There was a dignity to that somehow; to a life well lived and a real sense of "evergreen" and the continuity of years holding its own place in the ecosystem of this hill. Most likely, no one had ever specifically noticed it, and yet its life was of equal importance to that of any other individual that makes up this earthly ecosystem. What a lucky tree it was.

Those late June days saw the plant life along the roadsides grow quickly; the tall larkspur began to extend their racemes, and the first blossoms of cow parsnip opened. Numerous buds on the raspberries had me hoping there'd be plenty to harvest. Footprints covered the roads—no bear tracks, but plenty of whitetail tracks, including those of tiny fawns, and other large ungulates.

People so often think being out in the forest alone could be boring. But oh! It could not possibly be. Special moments are spontaneous, and that in itself keeps me on my toes at all times. To miss such unexpected gifts would be thoughtless. If we cannot accept such unencumbered magical moments, then why were we given eyes and ears? One must be constantly vigilant and receptive for the gifts of nature. I once spied a little bird while I was on my way up the tower. He was sitting on the very top of one of the spruces right alongside, and what a stunning blue coiffure this little fellow had! Putting several clues together, I have decided that he was a lazuli bunting. How often do we look at these seemingly exotic species in a guidebook, and wonder if one will ever present itself? Waving back and forth from the top of the tree, there he was, with all of his iridescent colours catching the sun's rays and reflecting them back to me. Would I trade such a moment for an evening of television? Never!

The generator, like a sewing machine on steroids, sawed away as I wrote, but blessedly it was breezy enough that I could still hear the trees rustling as counterpoint while the sun made its way to the western horizon. It had been a cool, changeable day in the foothills. The patter of rain on the lid of the rain barrel just outside my bedroom window, joined by the sound of trickling water in the downspouts, had roused me in the early morning hours. It had changed every hour or so from showers, to downpours, thunder, lightning, hail, sun and rainbows. It seemed I was the lucky one on this early July day with an overnight low temperature of only 5.5°C (it had only risen to 10°C that day), as Stenhope called in 1°C and light snow and Helios said that she was shovelling herself out. On Canada Day, Patrick had thrown the last snowball off Whisky-jack Mountain as part of his traditional celebrations. Mountain life is certainly unique.

That same weather kept the quadders away; I hadn't a visitor all day. In fact, I hadn't had anyone arrive for the better part of the week except for my supervisor and a few helpers on service day. Ah, meal choices! Current newspapers! Mail! I sent herbs and rhubarb up to Bill on Whisky-jack and wildflowers in to the office and warehouse so others could share the beauty of this ridge.

I could complete mundane and personal chores without interruption—I even changed the oil in the generator, managing not to spill oil everywhere in the process (it is an evil fluid). Since it was going to be five months between visits to a proper hairdresser, I had been practising the fine art of trimming my own locks. Even though I had my hair taken down as short as possible before I left, it had already needed trimming three times.

Busy Times

With the advent of July, the meadow surrounding Luna's helipad was an explosion of colour. I have a memory of a movie about Heidi, the little Swiss orphan, and an opening scene with her sitting in an alpine meadow on a sunny, breezy day surrounded by flowers. My meadow reminded me exactly of that scene. The western wood lilies had opened. Who needed fireworks to celebrate Canada Day? They were so incredibly vibrant, changing subtly every day from true orange to deep salmon. So much Indian paintbrush! At least a dozen different shades from pale yellow to peach to pink. The purple vetch accentuated the other colours, and everywhere were wild white geraniums. I did eventually find a few orchids of the ladies'-tresses variety tucked here and there; soft pinks were added by wild roses; the tiny mouse-eared chickweed had just begun to bloom and mats of pink pussytoes waved in the breezes from their spot on the crest of the ridge. Down the laneway to the gate, wild lily of the valley peeked out from beneath the more populous and taller species; cow parsnips were well into their flowering. The tall larkspur down the road was almost in bloom.

Not only was the week botanically opulent, but it had also been filled with wildlife. The elusive pileated woodpecker finally honoured us by swooping past as we walked down the laneway, finishing his performance with a few minutes' worth of percussion and closing with his haunting laugh. The number of tree-creeping species here was rivalled only by what I experienced while in Iowa. In addition to the pileated, there were hairy woodpeckers, sapsuckers and northern flickers. The dogs and I found the robins' nest in a young spruce tree at the edge of the road, tucked just perfectly so that I couldn't find an angle from which to take a picture. I watched a varied thrush search for his breakfast at the edge of the yard as I enjoyed a coffee on the porch one morning, and we scared a grouse as we all intersected at the gate one afternoon. Each day the Audubon variety of the yellow-rumped warblers (but not the myrtles) came to dine on the fluffy white dandelion heads that covered the lawn. On yet another morning, as I made my first trip out the door, I surprised a whitetail doe, no doubt one of The Girls, not six metres away. One afternoon she and I got into a staring contest, where she would peek at me over the tall underbrush down slope a bit, then stamp her feet and blow at me, only to run off to a new spot and stare again. The elk stags continued to graze on the cutblock down the hill from the helipad.

Recent showers gave way to sun, and the towers went back to either high or extreme hazard level. It was a tiring and frustrating week for so many. With the indices rising and news of fires in other districts, most or all of the towerfolk, including me, had called in smokes that turned out to be flarestacks, campfires, and so on. It's difficult to be at extreme hazard for days on end, knowing the chance of a fire is high, so you tend to jump at little things, and being wrong only makes it more frustrating. Daisy, who has been at her tower for many seasons, called in a flarestack that had apparently been there for years. She grumbled incessantly for the next day and a half, complaining that it wasn't marked on any of her maps and that she hadn't been told about it.

With everything so dry, there was plenty of road dust complicating the view. In one instance, road dust had come up and I quickly dismissed it. A glance back in that direction mere minutes later showed me bright orange flashes among the trees. My gosh, I thought, how could I have been so wrong? It had behaved exactly as road dust should! I grabbed my binoculars to discover the orange flash was the roof light of a grader.

Smoke from prescribed burns in Banff National Park reduced the visibility as well, some days changing almost hourly. Visibility had been bad for the past few weeks, but a wind shift could clear the air for short periods. As details came back into view, I pondered every flarestack smoke again, since it had been so long since I'd last seen them. Smoke and the angle of the sun and cloud shadows could change the appearance of the landscape dramatically. It would worsen as evening approached, the lengthening shadows and haze changing the view seemingly by the minute.

Patrick picked off a nice little late-day fire just on the west side of The Gate; Fred got a fire that ended up being a lightning holdover from a storm 10 days past.

And chasing all those smokes meant that the fire crews and helicopters were putting in long days and were tired as well. Since Luna was just a few minutes' flight time away from Shandray staging camp, a couple of helicopters had a bit of fun "buzzing" my tower on their way to and from the camp. It added a bit of fun to my days as we would wave to each other as they zoomed past the cupola windows.

Everyone was on extreme hazard level. I was just finishing up the last few chores on the ground before I made my climb up the tower for 09:00, when Larry from Stenhope radioed to say a particularly windy storm had just passed over his mountain and was coming my way. I quickly scooted up to the cupola to watch.

As it was coming toward me, the mare's-tail edges of the cloud encircled a very well-defined meringue-like middle. Between Whisky-jack Mountain Lookout and Luna, the smooth lower surface of the leading cloud developed a "belly button"—a dark indentation with some bubbling clouds in the middle. It never started to turn, but I watched it intently for the longest time. Larry had told me that I might see what he called "cold air funnels." As everything grew dark and the rain started to patter on the cupola, the winds hit in one sudden whoosh! Larry had reported them to be in the 75 k.p.h. range and I'm sure these were in that same category. The trees bent over, the rain came in torrents. Even the rain was fascinating, like hundreds of ghosts marching north; fine perfectly vertical sheets being pushed through the 9- to 12-metre trees that surrounded the tower. Not quite the "tattered curtain" effect since there was never the sensation of long strips waving in the wind, just vertical columns gliding through the forests. As the storm progressed and the clouds enveloped the tower, the mare's tails became even more beautiful and came to contrast perfectly with the distinct mounded centre. But no lightning was to be had from this storm; the heavens gave up a total of just one millimetre of rain and a couple of grumbles.

It was another oppressively hot and humid morning, with light showers here and there. Just as 19:00 ticked past, the towers and lookouts along the front range of the mountains, from Flattery Hill in the south and working their way north, began calling in first strikes. By the time I called my first strike it was past 20:00, and seeing as that made 11 straight hours in the cupola, with my last break four hours previous, I headed downstairs for a short spell to see to some personal needs. Helicopters were in the air to the south tracking down smoke messages; three fires were confirmed. I kept ears and eyes open while I tried to find a quick bite to eat. Considering messages that the radio, sky and forest were sending, I scooted up the tower for a peek, only to be met by a narrow, black boiling ridge of clouds less than two kilometres south of me, busily sending a great array of angry bolts into the side of the ridge, seemingly at my feet. At that point I retreated, having contacted Patrick on Whisky-jack Mountain to let him know that I was on my ladder with lightning in the area. My last check of the night at 22:48 showed that same storm trundling slowly northeastward and still sending out copious numbers of strikes. I prepared for what was likely to be a busy next day of watching for holdover fires.

These busy mid-July days brought their share of frustration. As the pressure mounted, just nothing seemed to go right. A couple of smoke messages I called in were less than accurate when it came to distance. Suffice to say I mentally beat myself up good, and at the same time reminded myself that if accuracy were important to Forestry, someone would have brought me a decent map by now. But it's nevertheless gut-wrenchingly frustrating when the helicopter is directly over the spot you gave them in your detection message, yet it isn't even close to the smoke. Added to those frustrations, Ddreena had come down with a nasty gastroenteritis as the temperatures started to soar, leading to additional worry about dehydration and flies. She looked like a wrung-out dishrag for a couple of days, but then rallied again, even taking a few moments to gallop around the yard to prove to me that she was fit.

New precipitation provided me with five straight days of low cupola occupancy, and I made the most of it. I mowed the lawns, weeded the garden, applied a fresh coat of paint to some roof trim and the outhouse (the latter much overdue) and then finally allowed myself just a bit of time to sit in the screened tent with iced tea at hand, listening to my radio, my dogs at my side. I even managed the occasional afternoon snooze between runs up the ladder.

Harvesting the garden officially began around this time. Mouth-watering spinach and baby lettuce salads, snow peas for my stir-fries, mounds of fresh basil, oregano and chives for my pizzas. By far, this foothills crop was the nicest lettuce I had ever managed to grow. A mouse also felt my spinach worthy, as I began to find only stumps where young plants had been the day before. I devised a web of flagging tape, bits of tin foil, flapping plastic lids and a bandana flag to scare it off, and except for the occasional incident, I managed to win the battle. The Asiatic lilies around the cabin had burst open to show a vivid rich orange beautifully accenting the yellow irises. The nasturtiums were due to bloom any day, another garnish for my salads. I kept pleading with the sweet peas to try to come along just a little bit faster.

Early one morning Ddreena began to woof out in the yard, and I found her staring off down the south-facing slope into the woods. I went out to see what had caught her attention—a bear! It was a lovely little black fellow, not all that big (probably just a yearling), and wonderfully healthy looking. I had noticed Ted occasionally going to the edge of the site and barking at something in the woods, especially between 16:00 and 18:00, and all this time I'd thought it was The Girls. Now I wondered how often it had been this bear. Also, Ddreena would often stop and take a long sniff at a particular spot on the lawn, so this fellow might have been passing through during the night. But it hadn't touched anything, and it certainly seemed shy enough, because I only saw it for a second before it scooted behind a bush. I suspected by the way it lurked from a distance that it didn't care for the dogs barking at it, and that was just fine with me.

There hadn't been many new feathered friends on the ridge in a while, but I was ear-to-ear delighted to have a golden crowned kinglet join our extended family. This species is another of those you thumb past so often in your guide and feel that you will never see. A bit bigger than I had imagined them to be, but so dainty, and so immensely happy, in much the same way that chickadees always seem to celebrate their existence. Ah, to be so glad just to be alive!

With birdsong filling the air, celebrating the joy of mere existence, the woods seemed to add visual harmonies by turning a vibrant fuchsia as the fireweed came into bloom, punctuated by the deep purple of the larkspur rising from the other waist-high wildflowers.

The dogs and I came upon a few strands of blue-eyed grass, and a clump of pink wintergreen just at the edge of the western cutblock.

There were other new flowers getting ready to bloom in the meadow to pick up where the wood lilies left off, and as the dogs and I would sit on the helipad in the evenings, I enjoyed the vivid flavour of the ever so tiny wild strawberries.

At Luna it always seemed to happen just at the end of the day: I would no sooner be down the ladder, with thoughts of dinner and two dogs wagging their tails at the bottom, than a storm would begin to roll in. I could hear the neighbouring towers begin to call in first strikes—sometimes it would be Whiteley and Stenhope first, sometimes Helios and Whisky-jack Mountain.

On this particular evening, it came in from the southwest and made its way between Brill Tower and Luna. The lightning strikes began on the ridges to the north, and I began to write detection messages as the smokes started to rise like crazy. Every time I finished taking a bearing from the firefinder, I would look up to find another smoke before I could even finish filling out the last message. I was jotting down bearings on a scrap of paper so I could quickly get the next, but losing track of it all as I tried to go back and estimate distance. As the rain began, scattering showers as it went, spooks rose from the woods and the sun sank low in the sky, casting a strange light upon the landscape. I watched some of the lightning strikes make their targeted trees explode into flames. The smokes were all so close together that the radio room declined further messages (the helicopters and crews would be able to see other smokes from the ones I had already reported). The adrenaline rush was incredible. The excitement got the better of me and I began over-chattering on the radio, and was promptly told to behave myself. Time for a few deep breaths and to sit on my hands. I could hear the towers to the east begin calling new starts as well; Drummond, Burrell, Lipton, Flattery Hill—the storm was following the foothills.

There were helicopters and tankers in the air everywhere. Crews were being divided and dropped at visible smokes. Between Brill and Luna the air was buzzing with air traffic. It seemed incredible that only two staff were in the radio room (the radio operator and the Duty Officer) and between them they could make so many decisions and direct so many crews and aircraft in such a short time. Two firefighters had been dropped just a few kilometres north of my tower and were fruitlessly wandering the woods seeking out a smoking tree. It wasn't until last light that a chopper finally came around and picked them back up, never having managed to find that tree.

By the time operations ended for the night, there were 22 confirmed starts and none that had gotten away. As always, a combination of skill and luck—and never take the luck part for granted. Those next few days saw plenty of extra air patrols and heightened awareness by the towerpeople as we kept watch for holdover fires.

Although it seemed hard to believe that Luna could hold any more beauty, the night skies also celebrated midsummer with the most brilliant orange half moon, the colour of which I had never seen outside of autumn before. The following night I awoke to long moonlit shadows in the yard. Even with the moon at only half power, the yard was bright with its glow. Mars was travelling progressively closer to Earth during this time. Millions of stars accompanied that huge flashing red planet those nights. From feather to flower to universe, all was well and good. There seemed to be gifts for everyone—the next day Patrick shared that he had been awakened in the middle of the night by an aurora.

It had been another nine days on either high or extreme hazard level, but finally 14.4 millimetres of rain dampened the forest. The DO had said that if we got enough rain we would be at low hazard level the next day, so I woke up with my wheels turning, determined to get laundry, lawn mowing and other chores under control. Since the effects of the rain upon the Fine Fuel Moisture Code (FFMC) would be gone within two to three days, I needed to be quick out of the gate to catch up. At 12:30, with laundry in the tub, the lawnmower out and four other projects lined up, the radio suddenly blurted out that we were now being put on moderate hazard level, and that we had to be in our cupolas for 13:00 hours.

I was furious. My FFMC had dropped to 43 (moderate hazard level indicated between 76–84), and I was expected to put everything away, finish my laundry, eat my lunch, prepare my weather report, and be in my tower with 30 minutes' notice. Somehow I felt I was missing the necessary phone booth in which to change into my super alter ego. But there was no time to enjoy a tantrum, and no one to listen (I must admit to venting to the radio-room operator, not that she deserved that).

Later that week during the time in the afternoon when fires were most likely to flare up, I had just taken a new well-site flare location from the radio room and was marking it on my map when I turned around to find smoke rising from a ridge about 10 kilometres to the northeast of me. I quickly radioed Patrick to see if he could confirm that it wasn't some strange road dust, and we were off from there. The momentum picked up as I collected the cross shot from Patrick, and the radio room, having overheard our conversation, had a crew and helicopter standing by before I could even get a pre-smoke message organized. It all went so well. It was confirmed to be a wildfire, and my estimate was accurate (as it should have been since it wasn't that far away and I had a cross shot), and the smoke was visible so the helicopter found it easily.

I cried. It had been so frustrating settling at Luna, between difficult terrain and uninformative and incorrect maps. When it finally worked, the sense of success was wonderful. The fire had started at the edge of a cutblock, even though from my perspective it looked like it was rising from a stand of mature trees. Despite burning mostly grass, it still took the crews until noon the next day to get the better of it because of how deeply it had burned into the duff and because of the lack of a suitable nearby water source.

Because the surrounding forests were so parched, with temperatures high and humidity low, they were doing prescribed burns in Jasper National Park. The visibility in our area had been horrid. I'm not sure I would have seen a smoke at 15 kilometres distance from the tower. The towerpeople were curious about just how big that fire was getting, because our view to the north had all but disappeared. Fred radioed to suggest that we look up in the sky, explaining to us that the huge, anvil-shaped cloud we could see was a pyrocumulus cloud. Such clouds are the product of the large amount of water vapour released by the burning trees, which can in turn generate more lightning strikes and start more fires. It was huge, with fuzzy edges and a very interesting bottom surface, made even more so by the colours of the late-day sun. That surface reminded me of the ceiling of a cave with just enough water and minerals dripping through to leave little rounded mounds that would turn into stalactites with time.

Later in the evening, the smoke from the Jasper fire blew into Pineridge district. My visibility went down to practically zero. The sky was brown. My throat became scratchy and the air reeked of woodsmoke.

Those days also saw the Lost Creek Fire in the Crowsnest Pass reach 7,000 hectares. It was only six kilometres away from the towns of Blairmore and Hillcrest, and the residents were on one-hour alert to evacuate. From Luna, I watched tanker crews fly overhead on their way to the Pass, and listened to the radio messages to the crews being scheduled for export.

But on our ridge, seemingly oblivious to the smoky atmosphere, the creatures that called it home continued to captivate us. Mrs. Golden-Crowned Kinglet had been bringing the fledglings to the spruce tree directly outside my kitchen window in the mornings. What a fuss as she scooted around trying to make everyone happy!

The red-tailed hawks returned and soared along the ridge. I watched the three of them play together for the longest time as they celebrated their ability to fly. If they had been just a little closer, I'm sure I would have seen them smiling. The gentle winds allowed long graceful swoops, at the end of which they would be almost vertical, nose pointed to the heavens. They would fold their wings ever so gracefully around themselves, almost as if they were hugging themselves in glee, then free fall, then slowly but surely unfold their wings once again, gently pulling out of their stall, and ready themselves for their next swoop.

Just when I believed that all feathered friends who were due to arrive must have done so, a brilliant red bird appeared clinging to the top of a spruce tree right next to my tower. His song! Coming down from the tower at day's end was like marching down the aisle of Westminster Abbey with the organ playing! I got the field guide out immediately, and identified it as a white-winged crossbill. Such a rich song (under-described in the guide as "canary-like") from a fellow with a beak like that! He and his cohorts stayed on, in the evenings forming a choir that filled the north slope with their sweet choruses. Bright yellow evening grosbeaks also joined us around this time, and between the two species a spectacle of colour was created in the treetops.

The classic August thunderstorms finally arrived. The lightning was spectacular, with towering cumulus clouds, beautiful sunsets, few fires and precious precipitation. Over three nights I accumulated 17 millimetres of the wet stuff, which finally dropped the hazard levels. I was grateful for the chance to sit in the screen tent between jaunts up the tower, Ddreena at my side, and even enjoyed a bit of the live coverage of the Edmonton Folk Festival on the AM/FM radio. My hope for a bit more quality time with Ddreena before she left had finally been granted us.

An evening storm netted a start on a ridge to my northwest. A classic small plume of smoke started rising from a lightning strike; just one or two trees, but a lovely string-like smoke, about 20 metres above the forest floor, no problem for a helicopter to sight. As I radioed in the report, the edge of the rain front reached it. By the time the two helicopters arrived, I couldn't see through the precipitation to help direct them and they couldn't locate the smoke. One helicopter crew was very sweet though, reporting that even though they couldn't see it, they could definitely smell it. Mother Nature had won again.

Yet another late-day storm was particularly dramatic as it literally assailed the earth with hundreds of strikes of every possible form. Some were visible for seconds, pounding jolt after jolt down the same beam, cloud-to-cloud and cloud-to-ground, with every conceivable zigzag. I picked up the storm as it approached from the southwest, and as it sidled up to my west side, it turned and came right over me. The electricity literally zinged around the aluminum trim inside the cupola, crackling and sparking. The area within half a kilometre of the site received at least half a dozen strikes. I had run up the ladder so quickly to start tracking the storm that I forgot my handheld radio downstairs, but decided that I would wait it out, hoping there wouldn't be anything to radio in. But a smoke came up just six kilometres away, a slow-starting sputtering little thing that the rain was only a few hundred metres away from dousing. I decided not to have a repeat of the night before; I would wait this one out. Suddenly, whoosh! The entire tree went up in bright orange flames, and the smoke column grew in size. I went charging downstairs to grab the radio, but by the time I got back up there was nothing to see. I continued to keep an eye on that spot for another hour or so until it got dark, and faithfully checked throughout the following days, but all was well. It never rekindled.

Ddreena developed a cough, and I feared it was congestive heart failure. She rested comfortably, but I worried as her cough worsened and she started to fade. She began to turn down her meals, although a bit of coaxing had her eating small mouthfuls from my hand. The cough eventually passed and her energy increased, but her appetite never returned.

Canis Major

Wolfhounds have a life expectancy of only around 7 years, and seeing as Ddreena was 10 years old at the end of the first season, it seemed a fair presumption that the flight from Helios would be her last helicopter ride. But she was also the most stubborn wolfhound I have ever owned, and she set about to prove exactly that.

I had deeply appreciated having the help of dog-friendly associates in getting her on and off the helicopter. The yard at Luna was much drier and more useful, but the stairs into the cabin were more slippery and harder to navigate for such a senior. Each day found me getting her lined up at the base of the stairs while I supported her hips, taking the weight off so she could hop up them.

With more navigable roads, we included walks every day and scheduled them earlier as time marched on and the temperatures warmed. But there was no denying it, Ddreena was failing. It was obvious from my perspective that she was still in deep personal denial about it.

Luna was an entomological godsend, because the biting flies were much less populous than at Connaught. When the flies were at their worst I would light a mosquito coil and put a blanket for her in my screened tent where she could be more comfortable when I was in the tower. In her stubborn way she had made it clear that being in the cabin during the day was not an option.

Her failing appetite was a lesson in innovation, because all food-stuffs were in somewhat limited quantity. Boiled eggs kept her happy for a while, but I ran out of those. Then it was cheese and ground chicken. One day a helicopter stopped in, and a smiling young fire-fighter handed me a current newspaper and a serving of pudding. Without a second thought, I gave the pudding to Ddreena.

It wasn't that I had high or unrealistic expectations. I knew the inevitable was coming, and I wasn't trying to delay it. I was trying to make her as comfortable as possible until she decided that she didn't want to go on any longer, and she hadn't come to that crossroads yet.

All this was happening at the peak of the summer's heat. The haz-ard level had been high or extreme for at least a couple of weeks, and it tore at my heart to look out of my cupola window at her sleeping on the lawn. I lamented that I couldn't spend a bit more time with her ... stroking her neck ... sharing last thoughts with her.

I decided I couldn't take the chance that she might pass away on some hot afternoon, when I wouldn't be able to get out of my tower to dispose of her until the end of the day. That situation seemed to be begging for a bear to arrive. So one evening, after 11 long hours in the cupola, I called over to Whisky-jack Mountain and asked Patrick if he would intervene at the radios if someone was trying to call me, and I went out and dug her grave. I had decided on a spot away from the cabin, on the far side of the tower, a place where the birds always seemed to sing, the breezes always rustled the trees, and the fragrance of the meadow flowers drifted by.

My level of anxiety with the situation had much to do with all the visitors at Luna. Since this was also peak camping season, the arrivals were relentless: four or five days out of every week and often in large groups. They were also relentless in their level of thoughtlessness. It was a constant barrage of, "What's wrong with your dog?" and "What's that hole for?" The strength it took to keep my chin up and square, to look people directly in the eye and in my most professional voice admit she was close to death, left me a limp rag by the end of those days. I longed for the privacy of Connaught. That pathetic look from the subpopulation of narrow-minded "juss take'er around back and shoot'er" contingent, implying that I was torturing her, was unbearable.

By month's end I could see that Ddreena was tiring, and her appetite continued to wane. Every day we spent a few quality minutes together, composed of quiet conversation and gazing eye to eye: is it time yet? Each day, the answer was no. Finally came that moment when she decided it was her time.

What the veterinarian had been able to give me wasn't the drug I was most familiar with when I had worked in a veterinary clinic. This substitute was not as popular but perfectly capable of carrying out what had to be done.

For days I had thought back to when, several years after leaving the clinic to move on to a new career, I had received a phone call from the Doc asking me to euthanize the hospital cat for him. Here was a man whose father had been a veterinarian, who had taken up the same career and had practised for over 30 years, but who couldn't bring himself to inject his own cat. I was taken aback by his sensitivity and honoured to be asked.

Now, however, it was my turn, and there was no one I could call. As with every pet I had ever owned, I had made the pledge to be a good companion, and when the time came, to end its suffering far more humanely than we allow for fellow humans. It was time for me to live up to my pledge. So one morning, before I could be overrun with more insensitive strangers, I prepared the syringe and steeled myself to perform the task at hand. I walked Ddreena over to the part of the yard where I had dug the hole, made her comfortable, held her close and made sure she knew how much I would miss her. I gave her the injection, and waited. Other than looking uncomfortable for a few minutes, she didn't even lose consciousness.

I started to cry. To have come so far, to have made the commitment and then to have failed. I had brought so many pets in the past to comfortable, easy ends. I couldn't believe that I would be dealt this card for my own pet. Ten minutes later she ate better than she had in weeks.

I was in shock. I was numb. I was past the point where I could even cry. Not only had I brought discomfort to my own pet, I was out of options. I had no other way to bring about a humane end as I had promised her. It was another three weeks before mail would be brought, and there wasn't even personal space to grieve, to think or to breathe. It was time to go up to the tower.

When I was finally able to get my brain in motion again, I called an out-of-province friend who was able to ship me an alternative. But that would only arrive with the mail. There were miles to go before the dawn.

Every time in those next few days that I sat down with Ddreena, the tears and the apologies flowed. I cried in my cupola. She remained stoic and seemed to accept what had happened far better than I. She managed the occasional tidbit and, thank heavens, she was still drinking plenty of water. By the end of the week, I noticed something happening on her forearm, around the injection site. The next phase of guilt began: the injection had been an irritant, and tissue was beginning to slough.

I, armed with bandages and ointments, and Ddreena, armed with determination, started into our twice-daily bandaging routine. Visitors to Luna came armed with a new series of thoughtless questions: "Is your dog's leg broken?" (How they thought that from the one little piece of terrycloth around her foreleg always amazed me.) "What's wrong with your dog's leg?" "What's that hole for over there?" It was endless.

I had a strange sense of relief and trepidation when the mail finally arrived. Being optionless was far worse than having a solution, no matter how demanding that solution was.

It was three days later before all the right pieces came together: Ddreena's final acceptance that she was ready and privacy in which to spend our last moments together. It was a beautiful Sunday. I had spent a bit of time that morning reading peacefully as she slept beside me on the lawn. She had wonderful wolfhound dreams as she chased rabbits and cavorted with other wolfhounds. I hadn't seen her have such a quality rest in days. When she woke up she turned to me, and as she set her gaze upon me, it was all there. She was tired, so very tired, and it was time to go.

And it happened. It was all so very right. I stroked her head and whispered to her, and she slept peacefully away.

Within the hour she was given back to nature and the universe that created her, with warm fragrant breezes floating past her resting place and birds singing overhead.

It wasn't until that moment that I realized Ted hadn't been a part of all this. I had taken myself to such a personal place to get all of this accomplished, that I hadn't noticed his absence. I went up to the cabin and let him out. He immediately went to Ddreena's favourite sleeping spot to see where she was. He looked around for her, and then looked to me. I suggested that it might be suppertime; his focus changed and we left that moment behind.

After supper I took him out to the gravesite. He gave the spot a serious sniffing over, then ceremoniously turned and defecated on the edge of the rock pile. I pondered this, because there was perhaps more than one possible translation. One was what I would suppose to be the more obvious and humorous: the "Well, that's what I think of you!" statement. I have chosen, rather, to believe the other: that he claimed that spot as ours, and made it clear to the forest that she was still part of our pack; she still resided in our territory. He wasn't ready to give her away yet.

There was such a huge hole left where she had been, with all the care that I'd extended to my faithful comrade in her senior years.

Ted bids farewell to Ddreena.

Ted and I filled time those next few weeks as we went out for our walks and returned with bags of stones to place on her grave.

Practicality and responsibility had taken over once again. I didn't want to leave any future towerfolk with a mess in case the frost decided to try to raise her. So she was firmly tucked in, under a grand mound of beautiful quartzite stones, worn smooth by the hands of the glaciers that had passed this way and laid them down on our ridge centuries ago. We added a few feathers to aid her travels and the gift of a plaque from our good friend the breeder, before we bid her goodbye at season's end.

The Trees Whisper Autumn Thoughts

There had been several interesting requests of the towers and look-outs of Pineridge Forest during those days that saw the summer vacation season coming to an end, with the ensuing final rush of tourists and sporting types to the mountains. On two occasions the more western lookouts were asked to watch for missing groups of mountain bikers (both eventually turned up safe and sound), as well as a separate request one evening for a helicopter and a couple of firefighters armed with chainsaws. Someone had done a good job of driving their truck off the road, pinning themselves in their truck wedged between trees in such a way that rescue crews couldn't access it. Once the fellow was rescued, the helicopter had to take him to the hospital in Rumseytown, because they didn't think he'd make it if the ambulance drove him into town. That quirky spell passed, and further requests for assistance with human emergencies subsided.

The first cool days were upon us, along with warnings of frost. I spent several chilly days in the cupola, finally giving up mid-afternoon most days and coming down for a hot tea to keep me going. Rain gave me yet another couple of days on the ground, but the hazard level still required hourly observations. Comfort foods such as a large pot of soup made with many things harvested from the garden began to fill my menu.

One lovely sunny afternoon I caught the tail end of a radio transmission from Bird Dog 6 saying that he was presently over Luna Tower. I rushed outside, heard nothing at first, but soon began to hear the drone of the engine. He swooped down over the site and I gave him a wave, thinking that he was just passing by. But he continued to circle the area, so I quickly grabbed my gloves and raced up the ladder. The bird dog, a Turbo Commander 690, swooped, circled and flew right by my tower, turning on its side, then pulling out straight toward the heavens. Then a lower drone began to rumble in the distance, announcing the arrival of three tankers.

Thus tanker practice began, and I had front row seats. A cutblock just to my northeast and a bit below my ridge was used first, each tanker making several passes at a spot marked out by the bird dog. When the tankers were empty, they returned to base to reload, leaving the bird dog to his acrobatics in their absence. When one by one the tankers reappeared, they proceeded to a different cutblock

The bomber drops its load as per information passed on by the bird dog.

that extended southwest from the helipad. I swear I could have put my hand out the cupola windows and touched them as they passed by. During most drops the tankers pulled up afterward, but they had a bit of fun on one run as they crested the trees, then quickly dropped their noses and dove down a ridge and out of sight. One tanker pilot even waved as she was going by. At the end of their practice, I radioed the bird dog to thank them and he replied that they might return in a week or so.

The weather warmed for Labour Day weekend, just in time for the campers' last hurrah. One fellow told of seeing more than 35 quads at the gravel pit (a spot new to me) and more than 45 quads at The Corrals, close to where the logging road crossed the Jean River. I compared notes with Patrick later in the day over the radio. He was not used to having more than a half dozen visitors a season, but with some trails closed to the west of his site, people were defaulting to his mountain. It turned out his day had been similar to mine, including a fellow who managed to drive his all-terrain vehicle up a steep talus slope to his cabin, rocks being thrown everywhere. I quietly thanked myself for the bit of personal quiet time I had allowed myself earlier in the week as I watched so many quads come and go from the parking lot.

There were four different types of groups that visited. Some were simply pleasant, polite and friendly, others acted as though access to the site was their right but totally denied my existence. The third group thought that the site and my time were provided purely for their personal entertainment, and the last had just lost their way and weren't quite sure what decorum was suitable for a fire tower.

I had to chuckle to myself over one group of visitors. They were of the denial type. They came up the laneway with several coolers and kids in tow, and proceeded to pick a spot in the yard not far from the cabin to have their lunch. Why, I wondered, would they sit on the ground when there were benches, plenty of shade and more privacy around the firepit? So I inquired, but they said they wanted to sit in that spot, and then they proceeded to turn their backs to me and eat their lunches. After their meal they had a photo shoot in the yard with one woman repeatedly waiting for my face to appear in the window, so that she could take pictures of me without even being polite enough to ask if I minded. I was greatly relieved when they finally packed it all up and headed back down the laneway.

The funny part? Well, that shady spot that they were so insistent to use for their lunch was, in fact, Ted's washroom area.

With the shortening days, Ted and I were able to spend more time out on the front porch taking in all the stars in the night sky. The satellites! I had a memory of Foxwood Inn in the Muskokas of Ontario, a group of us sitting out on the dock at night to watch one of the early NASA shuttles go by and remarking how amazing it was to be able to see a light cruise across the sky from such a distance. Now there were so many little lights cutting straight paths across the night sky. The waxing moon grew larger and a deeper shade of orange with every passing night. The auroras persisted and grew brighter. Mars twinkled alternately red and white, with the Milky Way like the silver cousin of the yellow brick road, and the entire panorama dusted with a million stars.

Even after the holiday, crowds continued to arrive on the weekends. I chuckled to myself as I wondered whether I would ever find solitude here. One precious moment saw an older fellow—sweaty, dirty, hair uncombed, shirtless, his pot belly drooped over his belt, his underwear sticking out from the back, his gait that stiff-kneed waddle that some heavy older men seem to get—make it to the end of the laneway. He set his hands on his hips, and with what I presumed to be his grandchildren around him and his wife trying in vain to ignore him, bellowed up to me, "So! Have you had sex up there yet?"

The forest embraced autumn as well; a scant few aspen branches began to turn gold, the wild sarsaparilla turned a vibrant maroon and gold, the fireweed tried to join in. The spruces took on a somewhat dusty and rust-coloured appearance as their cones turned from green to brown, leaving only the pines still dressed in bright green. The Bohemian waxwings came regularly, harvesting fruit from the meadow to the north of the tower. I was never sure whether they were taking the berries from the false Solomon's seal, the red baneberry, or the clasping-leaved twisted-stalk, but several would perch on the tower while one individual would do the collecting. They would then "gut" the seeds on the horizontal rails, leaving red juices and yellow seeds behind to decorate the tower. I laughed as I looked out the cupola windows and watched as they tried to turn their heads beyond what would seem comfortable to see me above them. The songs of the white-winged crossbills still trilled from the treetops; their red bodies glowed as they bobbed back and forth in the breezes. A lone blue jay flew through, a rarity here in the west. The raptors appeared more often, practising their soaring, perhaps getting ready for their migration, with species from kestrels to harriers to red tails riding the updrafts of our ridge.

Ted adapted a Frisbee for use as a water bowl. Each rainfall would see him dash out to enjoy the fresh water caught in its semi-permanent belly-up position. The squirrel made use of it as well, and would come right up onto the porch to help itself. It also started a daily ritual of climbing to the top branches of the spruce at the northeast corner of the cabin and "thinning" the cones each day. I was never sure what this exercise achieved. I could only see a very tense little squirrel working out its anxieties. Those tossed cones would travel surprising distances before hitting the ground. It sounded like a hailstorm inside the cabin as the cones bounced off the roof.

A radio transmission around this time carried the closing message to some towers in the district north of us, announcing tentative dates of September 23 and 30. The changing season had allowed bittersweet thoughts of the impending end of the fire season to enter my head, and now the reality was settling in. Although Vantage Peak Lookout had closed on Labour Day weekend the previous year when the towerperson returned to school, this year saw them put a sub in there, so I had a vague hope that detection was still necessary. My drought code had been increasing as of late, now sitting at 368. I was sure that it was still in the 500s between the mountain ranges. Of course, one good rainstorm could change everything.

A full propane tank was delivered to Luna around this time, and it was fun to assist again. The radio bleeped, giving me a scant few minutes to get out of the cupola and do the ground work. I hadn't even moved the empty tank off the little deck, so I rocked it quickly (even empty it weighs much more than I do) over to the edge, just in time for the helicopter to show up and drop the new one in place. The empty tank was connected while the helicopter hovered above, and he was on his way once again.

Time marched on. The rate of change was increasing on the ground, in the trees and in the air. The poplar stands through which the laneway traversed to the parking lot were now carpeted in the gold and red leaves of the understorey plants. With the white trunks of the aspens still crowned with their green leaves, it was a perfect autumnal mix of colour. The bird populations were in flux. The chickadees were impassioned; with just a few "chick-a-dee-dee-dees" on my part I would have an entire choir responding from the surrounding trees. I so dearly loved their songs filling my days. We startled flocks of white-throated sparrows in the low shrubs on either side of the road as we took our walks. If I stood quietly and let them settle, a quiet game of hide-and-seek could be had among the leaves. But the moment I moved, the sounds of them scattering were everywhere.

Cold, wet days came and went, after which it would get warm and sunny enough for the fireweed to release seeds again. Everything would become fuzzy with them, from the African daisies in the bird bath to the sweet peas that were finally making a concerted effort to bloom.

Low hazard days allowed me to get a few of the closing chores out of the way before that inevitable phone call announcing the end of the season. I mowed the entire site, including the meadow, which was a big job. I was careful not to push the lawnmower over the giant anthills, but to go out later with the hedge clippers and trim down the tall grasses growing from them. One hill was so big that I couldn't reach the middle of it.

With the mowing completed, I cleaned the lawnmower and sent it off with the helicopter on the next service day. This was the first time the crew didn't just throw everything in the door of the cabin and dash. The three fellows all came in and shared hot apple cider and some zucchini bread, which was a pleasant and welcome break. I was thankful the groceries and supplies arrived when they did, because the next day we ended up cabin-bound for the entire day as rain and cold winds gusting up to 58 k.p.h. tossed the trees about.

Hunting season approached. One walk that Ted and I took down the road had us come upon two sets of bootprints. The strange part was that they just suddenly began with no indication of how these people had got this far into the woods. They only travelled in one direction (away from the tower) and then ended as suddenly as they started. Along the same stretch of road were the tracks of a large ungulate, either an elk or moose. It had rained quite hard for the previous two days, so these footprints had been left since the rain. There were no quad tracks to go with them, which I took to mean that these fellows had been around since before the rain, because no one walks this far in. The last Sunday I had thought I heard a gun go off, and together with these footprints I couldn't help but wonder if I didn't have hunters lurking about.

Helios, Whisky-jack and Stenhope lookouts all received their share of snow during this time (although it didn't stay for long), and I could see fresh snow on the peaks behind Stenhope. Although I hadn't had a frost yet, I laid out some tarps and plastic sheeting beside the garden to be ready.

A large family of 11 surprised us with a visit one day. They were very polite and we had a nice chat. The skies were kind enough to clear and give them a nice view from the helipad.

I was on the ground when Brill Tower spotted a smoke. Bill radioed over to ask if I could see the smoke out by the Canal. I dashed up the tower and gave him the cross shot. It was a funny smoke, too big a base to be a flarestack, yet very dark coloured, typical of a flarestack, atypical of an average forest fire. It would grow larger, then sputter a bit, then begin to chug again. But it didn't really seem to be spreading, leaving Bill, Patrick and I all wondering what was burning. Bill decided that it must be a building, and he was right; it was a house fire. Local fire departments were left to act on it.

As the equinox approached, I sat cosily up on my ridge with a month's worth of food, but with a drought code of only 317 and forecasts of more wet weather approaching, while hearing from other towers the news that the Long Lake district had closed their towers a bit early and that a northwest district had extended their towers by two weeks.

As for us, who knew?

Three hunters arrived one afternoon. They came armed not only with guns, but with their best smiles on. They were pleasant and chatty, and I answered their questions about the tower and tower life until finally the inevitable question came out: So what wildlife had I seen lately? After receiving my standard answer of, "I'm sorry, I don't discuss my wildlife," the smiles dropped, they exchanged glances, turned on their heels and left, grumbling among themselves.

Later that week the first snow arrived, totalling 13.5 centimetres. I quickly tried to cover the garden plants that I hoped would be the most likely to survive the cold, using clothespins to hang sheets of newspaper over the sweet peas, applying plastic tarps over the rows of lettuce. Even though it got down as low as -4.5°C over those days, only two sweet pea vines got a touch of frostbite. The violas, lettuce, Swiss chard, spinach and parsley continued on, although I lost the nasturtiums.

Ted, of course, revelled in the snow, and I couldn't have had a brighter ray of sunshine on those gloomy days. The first morning the snow was just too wet and heavy, and try as he might to celebrate in full terrier glee, he bogged down with snowballs. The next morning the snow was drier, so he bounced himself to exhaustion. We took several walks down the road to enjoy the splendour of the trees draped in their white shawls, but we couldn't get far, because the young aspens were bent over the road from the weight they carried, and it wasn't cold enough to freeze the water sitting in the low spots.

The birds were a grand mix of sheer happiness and confusion. The chickadees, of course, were blissful with this sudden turn of events. A beautiful fat and happy robin sat just outside the window on that first morning with a terribly confused look on its face. The white-throated sparrows took it in quiet stride, while the waxwing fledglings were most perplexed by the sudden disappearance of their berry crop. But the best moment was when I was only 30 centimetres or so away from two chickadees and a kinglet as they foraged as a group through a spruce. Their usually exuberant chatter was muffled and reduced by the snow to contented whispers as they searched for their food.

I had hoped that Ted and I would have that entire snowy day to ourselves, but by lunch there were two fellows standing in the yard. Because of that same muffling effect, I hadn't heard the quad come in. It wasn't until the one fellow was kicking the snow off his boots against the front steps that I realized there were people about. They were just as surprised to see me! Drastically underdressed, wet and cold, the one fellow was shaking so hard I was concerned he was becoming hypothermic. So I invited them in. They hung their jackets over the heater, and I got a cup of hot apple cider into each of them. They stayed for about an hour, then were on their way again.

That next morning I rose from my bed to an unforgettable view out of my bedroom window. The sky was pink from the rising sun, and the snow draping the trees was the same gorgeous colour. The two huge spruce just beyond my tower framed the scene between them the same way that heavy velvet curtains frame a stage, and between them the twin does were peacefully grazing in the clearing beyond against a backdrop of rose-hued spruce. The scene was Christmas-card perfect.

Before long, however, it warmed again, and most of the snow melted away. It was much fun watching chunks of slush fall from the trees and getting out to sit on the porch in the sunshine again. The air was so fresh and moist, the sky so blue, the forest so clean with its carpet of white.

Patrick surprised us all by calling in a smoke mid-afternoon. From under all that snow? It seemed hard to believe. And at a distance of around 35 kilometres from his tower!

Snow-covered spruce at dawn.

There had been many fire permits radioed to the towers at that time, what with new seismic lines being cut and roadsides being cleared and the resulting brush being burned. There was a permit close to the spot where Patrick saw this little smoke coming up, but even though neighbouring towers and the office kept questioning him, he kept insisting his smoke wasn't related to the permit.

They sent a fellow to investigate by truck, with Patrick explaining how the spot was north of this, east of that, next to that cutline—doggedly determined to guide the fellow to it. It is by far easier to direct a helicopter with its aerial perspective than it is to guide a truck that is only staring at a wall of trees. By then the smoke had become intermittent. It was dinnertime when the investigator finally stumbled onto it and discovered a slash pile that had been burned before the snowfall, but which hadn't been extinguished properly and had now sprung back to life. Even under those soggy conditions fire hazards still existed.

Between these events, I'd been filling time with puttering: getting more things cleaned and packed, waiting patiently for good weather to hang out the laundry, cleaning paint brushes, knitting, trying to get a bit of writing done. On the few occasions I actually went into the cupola (it was just as easy to do an observation from below the cupola, unless there was something specific I needed to look at with the binoculars) I would end up being drawn to the side where my embroidery hung, uncovering it to remember what it looked like. Gosh, it seemed so long since I had actually sat up there for any length of time.

There had still been no word about closing dates. When the radio message went out to Vantage Peak asking him to call the tower supervisor, we all held our breath. But later in the day another message went out confirming that the towerperson needed to send in his grocery order. So it seemed that we would be remaining for a bit longer yet, weather allowing.

There were so many special moments during those quiet days. Small, vivid momentary snapshots set in memory during exquisitely peaceful, personal moments. Sensory images, like the fragrance of chamomile as Ted and I walked down the road and trampled the pineappleweed; how The Girls stayed closer to the cabin now that the hunters were near; the echoed sound of the pileated woodpecker tapping from miles away on an absolutely still afternoon; the aspens now turning to gold. In addition, how good homemade baked macaroni and cheese tastes on cold snowy afternoons; how leaves blown by the wind can sound different from day to day; the strange chatter of white-breasted nut-hatches as they moved through the trees in small flocks (I had never seen them flock before); and how I longed to hear the elk bugle.

The days marched on. The autumnal equinox was upon us. Ted and I would make a special trip out to the helipad and spend a few moments appreciating all these things from this little part of the forest that was our home.

No matter how dull the light was on sullen and overcast days, the aspens that had now turned to gold literally shone, brightening the forest with their glow, and the changing light over the mountains made every glance to the south a unique piece of art. Raindrops occasionally splattered on the front porch off and on throughout the day, but not enough that the drops ever melded together. I could almost, but not quite, smell autumn on the air (it does not have its own unique and musky fragrance in Alberta like it does in Ontario); perhaps it was only the memory of autumns past.

Other days were warm and sunny and allowed for a few hours' knitting out on the front step. There were also long wonderful walks down the road where we continued our conversations with the chickadees, and stood quietly while we were looked over by querulous grouse. There were very few ungulate tracks on the road these days. Both the grey jay and robin families visited regularly to dine on the countless grasshoppers that still inhabited the lawn. Patrick reported treading softly around his lookout because a black bear with two cubs was enjoying the berries just below his cabin.

Twice I found a mouse floating in the bucket that sat under the rain barrel faucet (you can always make use of what escapes from a dripping faucet). They were so tiny that I was sure they were young just leaving the nest. I felt badly for their early demise, especially since they were being polite enough not to share my space in the cabin. In an attempt not to be wasteful, I put their poor little stiff bodies up on the tower beams. The ravens came along and made nutritious use of them. I made it part of my regular routine to empty the bucket at bedtime, and I put out a jar lid with a bit of water in it so they had a safe mouse-sized water dish of their own.

For all the hunters in the forest during those days, I had yet to see a sign that anyone had killed anything. In particular, earlier in the week I had a problem with a hunting group and went out on reconnaissance patrol to try and figure out if they had in fact shot anything. Although my patrol happily turned up nothing, I now found myself checking regularly for signs of hunting groups during our walks. I began to carry the handheld radio with me. It seemed so wrong to be a prisoner of conflicting lifestyles in a place so free.

The message finally came from the office that closings were near, and that the final schedule would be passed to us the following Monday. I cleaned all the windows, washed walls and cupboards, turned the gardens under, tidied the gen shed. I started packing up more non-essentials.

But by far the hardest thing to organize did not fit so neatly into boxes or luggage. My soul is never ready to leave. It seemed hard to believe that five months had flown by so quickly. More so this year than last year because I had become so totally comfortable with this lifestyle that I just didn't miss anything; there was nothing that I was in a hurry to move back to the city to regain.

After a cold, damp and dreary spell, the dawn sky was clear and the morning warmed both in temperature and light. The brilliant gold of the aspens was further enhanced by the sun shining through them, causing me to squint as I looked up at them. I climbed the tower and was met by the tikking of juncos, the nasal sounds of the nuthatches, the eternally happy chatter of the chickadees and the kinglets. The whisky-jack family added their assortment of sounds; the robin family sounded perplexed at what all the noise was about. This was not morning vespers, this was the Hallelujah chorus. Everyone of feather and wing joined together to sing the praises of the return of a warm and sunny world, and all on the north side of my ridge. I was reenergized from all the joy.

A long run of warmth and sunshine followed. The highs of around 11.5°C grew each day, reaching 24°C, with only the occasional day able to lay claim to a few wispy high clouds in an otherwise clear blue sky. The nights grew warmer too, allowing the bedroom window to be open again.

Evening scheds were moved to 16:30, so Ted and I would leave immediately after that for our daily walk. The grasses had reached the cured stage. The gold colours to which they first had changed now turned to brown, their seed heads matted, the seed scattered on the road from the white-throated sparrows feasting upon them. The thrumming of grouse could be heard. The nuthatches had curiously turned from the white-breasted to the rose-breasted variety; the yellow-rumped warblers were on their way south, and the hawks were absent as well. The alders were for the most part bare. Those leaves were brittle and crunchy, and very kickable. The forest floor that was once a motley mix of golds and reds was now mostly brown as well, but there were still little spots of colour here and there from a rebellious plant refusing to give in—the mauve of the occasional aster, and the red of the rosehips. The remaining fireweed seed was now mostly knotted into the curled pods; few became airborne any longer.

The night skies were just as vivid as the day's, clear and cloudless, filled with a million stars, sprinkled with shooting stars or draped with auroras and topped with a waxing crescent moon that began each night in full blush until it climbed above the mountains to my south. The Milky Way blazed its path directly above our heads; satellites carved their slow straight paths among the stars. I would go out to the helipad every night with a cup of tea around 22:00, with the last of the day's warmth still hugging the ridge.

From my tower, the landscape had changed to that of a patchwork quilt. The golden aspen stands were now so vividly different from their spruce and pine neighbours. It had taken a few days to get used to the new tones because light grey smokes do not stand out against gold the way they do against darker colours, and because with the resurgence of warm temperatures it had been hazy. The aspen stands that were the first to turn gold were now turning brown, and the leaves beginning to drop. I would look down at the yard and watch the leaves fall, flashes of gold among the tall white slender trunks catching the rays of light that had succeeded in making their way through the canopy, rocking gracefully in their descent.

The Girls still stopped by several times a week. They were changing colour too, from their vivid gold summer coat to the duller camel colour that would allow them easier disguise in the winter forest.

An ungulate of a different sort, a cow moose, came through our yard one afternoon at a full trot, creating a lot of noise in her wake. Whether she was being chased or could hear the voice of a handsome bull calling her name was never to be determined, but she continued on her mission across the middle of the next cutblock and disappeared over the edge of the ridge without breaking stride.

Mama Squirrel's boobs continued to get bigger and bigger, so I began to call her Bessie. What with all the heat and low humidity, she came several times a day to Ted's water bowl for long drinks.

Igor disappeared around this time, I guess to wherever it is spiders go for the winter. I thought she might return with the warm weather, but her garage stood empty for the rest of the season.

The gardens revived as well. A bit of hoeing and fertilizing and it looked like the second crop of spinach might make my plate for Thanksgiving dinner. There were still a few heads of lettuce and plenty of parsley and Swiss chard left; my sweet peas bloomed again; the few remaining nasturtiums rallied.

And so, with all of this hot, dry weather, my season was extended. My supervisor allowed me a quick food order and let me know that I would probably be open until after Thanksgiving. My FFMC was now only one small number away from extreme hazard level. There had been three smokes called in the past two days, and still no rain in the weather forecast. And who could say—in true Alberta fashion it could have snowed the next day. But until that time, I was trying my best to take in every lovely moment.

The Hunter Moon

It was Saturday night once again. Ted and I had returned from our walk; he had inhaled his dinner and was sleeping peacefully. It had been a cold, crisp day; we reported a mere 6.5°C at afternoon weather. The sky changed frequently throughout the day; at that moment there was a dark grey thick cloud front coming across the mountains, decapitating the peaks on its route northward. Larry and I concurred that it looked like it might bring snow; I made a mental note to remember to cover the gardens.

In the previous week temperatures had soared; this new week was the opposite. Indian summer slipped away, yet frosts remained few and far between. The daytime temperatures slid from 25°C to 13°C; the overnight temperatures from 13°C to 0°C. The trees that had rebelled against the short Alberta autumns, struggling to keep brilliant colours, were now bare, and it seemed to happen overnight. Now the view from the cupola showed a lawn turned gold, and the paths through the woods where we frequently walked were clearly visible through the bare branches.

It was a week of contrasts. A few more visitors arrived, but as far as I could tell no hunting group had stayed on our ridge. There had been a few odd episodes, like the weathered old fellow who arrived with his rifle on his shoulder. Until now, all the hunters who had visited had left their guns in the parking lot. But this fellow was to have none of that, keeping a firm grip on his rifle the entire time he was in the yard. As he walked out to the helipad, I silently prayed that The Girls wouldn't pick this moment to pop by for a visit. I had a funny feeling that the grouse feathers we had come across scattered on the road a half an hour's walk away might have been his supper that day.

The next day, as Ted and I headed home on our walk, I could hear a quad from behind a rise. As we crested that rise, I could see the young man driving the quad surveying the adjoining cutblock with his binoculars. As we weren't making much noise on the soft road, we ended up startling him, and his first response was to reach for his gun (although he never lifted it from its mount on the side of the quad, thank heavens).

I had a smoke one afternoon, a sweet little thing that I immediately imagined to be the perfect ending to a season of nothing missed. A typical woodsmoke, but rather lackadaisical, it came up from a ridge of mature spruce. I sat and watched it for a while to make sure it wasn't a campfire and was actually going to persist long enough to get a helicopter to come in and check it. I called Stenhope to see if he could spot it and help me with a cross shot, but to no avail. It was almost exactly where Larry had called in his first strike from a storm the previous evening, and it was too big to be a campfire. So I called it in, and the helicopter was around in short order, since there was one nearby on a regular patrol. It only took a couple of "No, a little more to your north" encouragements from me before the helicopter spotted it after it threw up another puff. I let out a gleeful, "Yesss!" as the helicopter radioed in to the office, "One mile back, smoke sighted, we'll have a quick assessment to you in a couple of minutes," for if they thought it was something from their short distance, then it must be so! But ah, all the difference a mile can make. It turned out to be coming from a hunting camp that was using a large tent that allows for a wood stove.

So when yet another smoke identical in behaviour came up just a few days later, about five kilometres off the edge of my ridge, I decided on a different approach. I radioed Larry just to let him know that I had it sighted and that I was keeping an eye on it, but indicated that I wasn't going to call it in until I felt there was a reason. Later that afternoon, a call came from a helicopter saying they were in the area and offering to check it out for me. I graciously accepted, and sure enough it was another of these same tents. It was a satisfying feeling that I hadn't wasted helicopter time by pulling it from the other side of the district for something that wasn't an emergency, and that I felt sure I could now identify this particular kind of smoke.

That week the nights were much less star-studded, due to the effects of the full moon. There was no Milky Way, no auroras or shooting stars. In their place were long shadows from the trees edging the yard and, of course, the tower; late dusk produced an incredibly picturesque, rich azure-blue horizon that highlighted the razor-sharp silhouette of peaks. The stars appeared as the sky darkened, seeming to descend to meet those peaks with the last wisp of daylight, the mountains and night sky becoming one.

The nights were now almost too long. It seemed senseless to get up in the dark in the morning, because it only used up more lamp oil and candles, but neither could I stay in bed that long. My mornings were lazy, including a bit of reading and knitting until I needed to make a run up the tower around 10:00. I spent from 14:00 to 16:30 each afternoon in the tower, as I tired of Bill's periodic cranky calls from Brill, "Can't you see that smoke?" And after all, although the office didn't assign us anything greater than low hazard level because the days were so short, the FFMC was still in the high hazard level, so it made me feel proactive and like I was earning my pay.

And so it was that Thanksgiving arrived. My holiday dinner was composed of the last canned ham (but glazed with a wonderful concoction of spices and dried cranberries) complemented with a salad and Swiss chard from my garden to give it that sense of harvest. As I sat down to my feast, I thought of all that I had to be thankful for: my second successful season at my dream job, my friends and family, Ddreena's last days spent in such a beautiful place, all the creatures who visited Luna, the beauty of nature that surrounded me and the space and time to enjoy it.

In Pineridge Forest, what were once 11 active towers were now 6. Usually by this time it would be down to 4. I was counting my blessings.

The cumulus clouds were the first to turn golden as they passed beneath the high wispy clouds that formed myriad sand-like undulating ridges. The clouds to the east gradually turned to salmon-pink; the ridges and mountains to the south turned shades of indigo and deep lavender. The trees to the west of where we sat on the porch eventually lost their detail, turning into leafless silhouettes against the brilliance of the sun's last rays. Among those bare branches, the whisky-jack family bullied the nuthatches; somewhere in the background a woodpecker tapped out its Morse code messages.

As the month rolled on, the inevitability of the closing of the tower occupied my mind more and more. The call finally came, informing me that they had made the decision for Luna to remain open and asking how my supplies were holding up. As it appeared that such decisions transpired on Mondays, I said I was fine for another week, and then waited to see if they would hire a helicopter to bring groceries in or transport us out. There had been only four millimetres of precipitation since the third week of September; things were thus a bit dry. The wonderful second coming of Indian summer persisted for several days with only a small chance of precipitation in the near future. I took out all the old weather record books (1970–2003) and worked out that October 11 was the average closing day for Luna.

Those next two days that it took for the warm weather to convince the cold weather to leave were strange, ominous days. Gold and pink colours remained in the sky until noon; the clouds were deep greys and looked grumpy. It became a greater challenge to detect smokes in the constantly changing light. Cloud shadows swung around and traipsed across the landscape; and the sun, now at a lower angle due to its southern course, moved quickly across the sky. Haze and smoke dulled the details of the landscape.

So one day when I was doing my final observation of the day just after 18:00, I caught my breath as I saw a light grey spot off behind a ridge. I radioed over to Brill Tower, but since it was now past the end of the tower day, I got no reply. I then radioed Drummond Tower, but Daisy was already down and had to climb up again—only to find that she couldn't see over to the area of interest. The light was terrible by that time, the last light of day turning everything distant into strange unnatural colours. And this smoke wouldn't rise, wouldn't puff, wouldn't move—common behaviour for that time of day, but it's those things that help identify smoke from other events that cloud the sky. I hesitated, as it would mean sending men out just 15 minutes before stand down. That was okay if it really was a fire; but I'd feel guilty dragging their day out for something I was only imagining. I made the decision to call it, applying a heavy dose of the "better safe than sorry" reasoning, took a deep breath and wrote up the detection message. I decided that if it was a smoke, it had to be the Evening Star Reservation's dump 35 kilometres away, and recommended that they begin their search there. Fortunately, the district's only helicopter had just dropped off a crew at Shandray firebase and was returning to Rumseytown airport, and therefore only required a short detour to check it out.

If you think you've seen some crazy dances in the end zones of football games, you had to see my fancy footwork around that cupola when the helicopter confirmed that it was the dump. I departed my cupola that night with a pretty big smile on my face.

Notwithstanding the dryness, the next week was relatively quiet wildfire-wise, although the number of fire permits read out over the radio each morning continued to get longer. New roads were being built into oil patch leases, generating huge piles of brush to be burned. Some days I had no fewer than seven huge smokes in every direction around my site, some behind ridges and mountains, some far on the horizon. It drove me crazy trying to keep an eye on them throughout the day. I would keep seeing them out of the corner of my eye and panicking before taking a better look and reconfirming the smoke was rising from a location covered by a permit. Bill had told me that visibility improved in October, which was confirmed when I radioed him to ask about a smoke on the horizon one afternoon. His reply was that it was the mill plant in Dunford 75 kilometres away, almost twice the distance of my primary area of responsibility.

I managed to coddle the garden through more cold days. The mule deer buck, now crowned with four points, and I startled each other one morning when I went out to do weather. He bounced away like an antelope only to show up again mid-afternoon standing in the laneway, peacefully staring at me. I tippy-toed into the cabin for my camera but by the time I returned he had moseyed off. Then at dinnertime, as I was talking to my mother on the phone, there he was standing just outside the window, nibbling the weeds from around the edge of the garden. I had to laugh when the next day I saw that my snow pea plants had disappeared; I guess he had dined on those as well! While I was watching this fine fellow, Fred, now on Burrell Lookout, shared with us over the radio that he had watched a cougar walk through his yard.

Visitors were now intermittent. The ridge seemed more at peace. A helicopter came by with a couple of fellows who installed the new fall-arrest system on the tower. They also brought along another tank of propane, allowing me more than enough fuel to end the season. Ted and I continued to take afternoon walks to enjoy the changing scenery. The skies were clear again and the temperatures moderate; evenings included trying to find the Andromeda galaxy in the night sky.

Although I would never admit to being ready to leave, I found myself thinking about the third concert of the season being presented by the folk music club at which I volunteered each winter. With so many talented volunteers on my team, I knew that they could run things smoothly in my absence. However, I had been looking forward to this evening of visiting with friends and enjoying music, as well as completing plans for a nice dinner and a concert in Calgary, now just a few days away, to celebrate the end of the season. It would come. If I had been in Calgary at this time, I knew it wouldn't be long before I would be thinking of how much I missed being out at the towers.

The ridge continued its preparations for winter. The changes were slow and subtle, since much of the forest had already gone to sleep, but if I looked closely enough, I could find them. I surprised The Girls one morning. Since I hadn't seen them in quite a few days, it was good to see them alive, happy and fleet of foot. The very next day as I was getting ready to exit the loo I peeked out the window to see a whitetail buck standing in the yard. Little twigs grew from his head; I figured he was a yearling, in beautiful condition.

From the cupola I watched pine grosbeaks as they rocked back and forth, clinging tightly to the tops of the spruce trees in the wind. I made the final decision on my resident raptor: it was a light phase immature rough-legged hawk. Another ornithological surprise this week was a boreal chickadee hanging out with a flock of black-capped chickadees. His song was just that little bit different, and although he seemed a bit shy initially, a bit of "call and response" brought him out on a branch just 30 centimetres or so away.

Larry on Stenhope was next to have a cougar saunter through his site. He had gone out at dusk for a short walk around the cabin, but stopped to look around when the hairs on the back of his neck stood up. When he turned around, the cougar was standing out by the buildings owned by the telephone company, a hundred metres or so away.

The office surprised me the next Monday by radioing before 09:00 and announcing that we were all remaining at our sites until the end of the month, following that with a request for quick grocery orders. The next day the helicopter came in and dropped off the order and fresh water, and was quickly on its way.

And that, noted gleefully, was the last human I saw for a week. Although our walks showed that there had been a quad scouting the area, I hadn't heard it. That was at least partly due to the fact that the week had for the greater part been very windy. There seemed to have been battles all week over which of two competing weather fronts was going to occupy this area. While some days were bearable on the ground, most days the temperatures hovered around 5°C, with winds in the 20–35 k.p.h. range and gusting up to 50+ k.p.h., leaving me feeling drained of all body heat after a run up the ladder. Thus it was a long week of being indoors, and I had to admit to a few bored moments. I wasn't the only one struck with cabin fever; Larry had taken to walking as well, no matter that up on his mountain it was probably five degrees colder and the winds 10–20 kilometres stronger.

In the past 30 days our ridge had received a mere 11.8 millimetres (barely half an inch) of rain.

No sooner were the supplies delivered, than I had to get the garden covered to protect it from a likely overnight frost. In my rush, I ended up setting the lantern down on a rough spot, and over it went, breaking the mantle and the chimney. Since it was now getting dark around 18:30, the thought of long evenings without sufficient light to read or knit was a bit unpropitious. Thank goodness it came to me later in my sleep where I had packed the spare mantles, and I awoke with a plan on how to glue the chimney back together. So all was well again in short order, although the chimney looked funny stuck together with silicone caulking. Other than changing colour, the silicone didn't seem to mind being heated each night.

I found the Andromeda galaxy during those days. There had

been few clear nights, but with the new moon, the clarity I needed to correlate my star guide was now possible. Andromeda required the binoculars to see it, but it was fascinating. I had heard on the radio that the light I saw had left Andromeda two million years ago. I thought about what the area around Luna might have looked like at the moment that photon left that far away place, and pondered what slow, steady processes might have sculpted what I saw today. From seabed to mountain peaks, it was a fascinating thought.

Goodbye Fair Luna, Goddess of the Moon

The helicopter was due to arrive at 11:00 the next morning: closing time had come yet again. Monday had brought news of an approaching storm, and the remaining tower and lookout observers were given the word to close down. Since Luna was the last fly-in tower open, a helicopter needed to be coordinated while the other towers and lookouts could just drive away.

I found myself reciting the words I had written for the Jordan newsletter and thinking about what this season had meant to me. So different was this year's experience at Luna compared to Connaught.

While I found that I hadn't nearly had my fill by closing day at Connaught, this six-month stay at Luna had been just right. With the forest now prepared to accept winter, it somehow seemed timely for our season to end. Ted and I took our last walk that afternoon. I harmonized our last song with the chickadees, and we stopped ever so quietly so I could absorb the sound of the wind in the pines. Other than the chickadees and Mama Squirrel, there wasn't anyone else around that day. When I thought of the last time we had startled The Girls, I was left with the good feeling that we had left them with exactly the right fear of humans; there was no need to worry about them once we left.

I made my last trip up to the cupola, stripped the firefinder of the scope, put the cover in place and packed the binoculars away. Parts of the garden had been turned under. I left the spinach for whatever little creature had nibbled it over the season along with the two largest heads of lettuce.

The cabin was in that "middle" state of packing again. I had tidied the desk around the computer and put the photos away; there was only one day's weather report left to add to the weather record book, and then it could be added to the archives.

While there had been many challenges and achievements at Luna, it had not filled my soul the way Connaught had. My observation skills had improved over the year. It had been very frustrating adapting to the landscape here, and I remembered tearfully pledging earlier in the season that I would rather stay than have to learn yet another area. But other things I now realized had a value to me as well, such as privacy and a need to feel some sense of team spirit. They weren't to be found on this beautiful ridge in the foothills. Although I told myself I could continue to ponder until it was time to re-apply, I somehow didn't envision myself returning. And that thought gave me a strange sad spot in my middle as I walked out to Ddreena's grave and reflected on her part in our summer here. I searched my soul and let her go; she belonged to the ridge now.

Larry left Shandray Mountain that morning. Bill left Brill that afternoon. So at afternoon weather and scheds I was the only tower left on Channel 3. It was an odd feeling. The sunset was painted in pastels with clouds telling that snow would be coming in over the mountains. I planned to be up in plenty of time in the morning to enjoy our last sunrise.

The tally at Luna for that year stood at 21 smokes:

- Fires: 8
- Flarestacks: 3
- Dump: 2

- Nothing found: 6 (three were lightning strikes extinguished by rain before the helicopter arrived)
- Previously reported: 1
- Campfire: 1

Knock on wood, none missed. The district total stood at 192 fires.

The next morning I woke up to 36 centimetres of snow and zero visibility. At weather, the greeting from the radio operator had changed from, "Good Morning, Towers!" to "Good Morning, Sharon!" I was the last one. The helicopter was placed on standby. The day was spent packing as best I could. It was tough to finish the packing until I knew when I would be leaving. Little by little, packed boxes started to be re-opened as things were needed. Ted and I revelled in the beauty and the silence. I had lots of propane and food, and we weren't in any hurry. Shovelling snow gave us a reason to play outdoors.

By that evening, there were 46 centimetres of snow. By the next morning there were 56, and still zero visibility. We shovelled some more.

By the time the snow ended, we had a grand 66 centimetres of snow. For three more days the skies were either clear over Rumseytown or over Luna, but never both. Finally, a small break in the clouds allowed for a quick decision to get in, and the helicopter was dispatched, arriving at Luna mid-afternoon. A path was dug to the helipad, and the loading began. The fellow sent in to help with the loading prepared to disconnect and pack the generator, but I stopped him. Because it would take two trips to get everything out, I couldn't let him shut off the propane at the tank to disconnect the gas line at the generator. If the skies closed again and they couldn't get back for the second load, then I wouldn't be able to heat the cabin or cook any food. The generator and I would just have to leave at the same time.

The skies allowed for the second trip to finish the closing, but it was by no means an easy day. The windows were frozen, making it

Sixty-six centimetres of snow have blanketed Luna by the time the helicopter is able to pick us up at the end of the season.

difficult to get the window boards up, and snow was tracked back into the cabin faster than it could be cleaned up. It was even colder back in Rumseytown, where we unloaded the last of my gear well after dark. I fell into my bunk at the bunkhouse that night with the sound of television emanating from the common room and traffic sounds coming through the windows. It was back to civilization for Ted and I again.

"*This is XMG 722, down for the season.*"

**Luna Tower
Incomplete Species List
42-12 W5M
Elevation 1,417 metres
May–October, 2003**

Alder

Aspen

Asters (white and mauve)
Asteraceae spp.

Baneberry, red and white
Actaea rubra

Black spruce

Clasping-leaved twisted-stalk
Streptopus amplexifolius

Common paintbrush
Castilleja spp.

Common pink wintergreen
Pyrola asarifolia

Common yarrow
Achillea millefolium

Cow parsnip
Heracleum lanatum

Cream-coloured vetchling
Lathyrus ochroleucus

False Solomon's seal
Smilacina racemosa

Fireweed
Epilobium angustifolium

Fleabane (white)
Erigeron sp.

Grasses and sedges

Heart-leaved arnica
Arnica cordifolia

Hooded ladies'-tresses
Spiranthes romanzoffiana

Lodgepole pine

Mosses

Mouse-ear chickweed
Cerastium arvense

Northern bedstraw
Galium boreale

Oak fern
Gymnocarpium dryopteris

Pineappleweed
Matricaria discoidea

Pink pussy-toes
Antennaria rosea

Purple clematis
Clematis occidentalis

Tall larkspur
Delphinium glaucum

Tall lungwort
Mertensia paniculata

Veiny meadow rue
Thalictrum venulosum

Western Canada violet
Viola canadensis

Western wood lily
Lilium philadelphicum

White spruce

Wild lily-of-the-valley
Maianthemum canadense

Wild rose
Rosa arvensis

Wild Sarsaparilla
Aralia nudicaulis

Wild vetch
Vicia Americana

Wild white geranium
Geranium richardsonii

The Weasel Hills: Tamarack Lookout

It felt strange to drive directly into a tower site instead of travelling to an airfield and loading a helicopter. But here I was, driving into Tamarack, my third new tower in three years. I would be starting over again, settling into another new cabin and learning the landscape of a new section of the forest over which I was assigned to watch.

I could see that this site was not going to fulfill me visually like Connaught or Luna had—but I could also feel that there was peace in this place, not like Luna where the energy only served to irritate and frustrate me. Life presents interesting balances, and it appeared that I might be practising the art of quiet acceptance yet again.

Ted and I piled out of the car and looked around. The cabin was old and worn out. A date of May 18/64 was inlaid into the concrete step. As I opened the door, the fragrance of the mothballs left by the previous towerman to deter rodents hit me. While it had worked, I wondered if I'd ever manage to get the smell out. The floor was the worst part. It had been a whole summer's work for a previous tower-man to scrape out the rotting smelly carpet, leaving behind tiles that were broken, loose, curled and decorated with the swirled remains of the old glue. But at least the walls had a fresh coat of paint and the windows were newer, with the screens intact. A new armchair awaited me. It would be the first season in which I would be able to relax in the cabin on something other than a kitchen chair.

My supervisor and I dragged the soiled, decrepit old couch out to his truck for disposal; an old wringer washer leaning against the side of the generator shed went as well. Just that little bit of initial purging left me with the feeling that perhaps there was hope.

The box of equipment from the warehouse contained my new full body harness; with this discovery came the acceptance that the new fall-arrest systems installed on the towers last summer were now operational. I sighed as I resigned myself to the reality that it was going to be a long summer getting in and out of that thing a half-dozen times a day. It wasn't as if we hadn't known it was coming; safety is such a big issue these days, and regulations stated that it would be mandatory. I closed my eyes and imagined hundreds of old towerfolk turning over in their graves at the thought of wearing such a thing. Added to the constraints of the harness was the news that the buddy system was now mandatory as well. Before the body harness and I would be going up or down that tower as a team, I would be radioing a neighbouring tower to state my intentions to climb, only to have to call again from my destination to claim success, with the whole district listening. Somehow the freedom that was the main course of soul food reaped from tower life was wilting on the plate. Even a life of solitude in the boreal forest wasn't far enough to escape from the complexities of modern life any longer.

Times they were a-changin', even in just three short years. I would need to look for solitude in new creative ways.

But other bits of life at the towers remained unchanged. I watched a pair of northern harriers circle the site before the end of the day. Once again I wondered (if I were to think in an Aboriginal spiritual way) if these graceful creatures might be my animal guides, because they had been my companions each season, playing a part in our adventures. It was good to be greeted by old friends.

And so began our third season.

"This is XMG 746, open for the season."

Nature as my Roommate

We tend to think globally these days, universalizing everything with some idea of commonality, and yet nature is anything but one colour uniformly applied across a canvas. Generalities do not apply. This northeastern side of the province proved itself unique from either of my other two tower assignments. I thought back to how Valerie and I had noted the differences even in the 40 kilometres between Stoney and Connaught towers.

I fretted much over the silence of the forests that early spring, but gradually, with the odd warm day, it became evident that I had arrived before the birds. A neighbouring towerman's thoughts on this concurred with mine during evening conversations.

But as April gave way to early May, with temperatures hovering only around the -5°C range, the Canada geese and sandhill cranes started to arrive. Warblers began to warble from the wood; large flocks of robins dined on my lawn, a stray Swainson's thrush joined his red-breasted cousins. A yellow-shafted common flicker added laughter from his perch on the tower and then moved to the cabin roof for a percussion solo beat out upon the chimney. Not to be outdone, a pair of ruby-crowned kinglets foraged from the willow tree just outside the cabin door. Even among the snowflakes, the male could find something to be happy about as he stretched his entire body skyward as if on tippy-toes, his scarlet crown erect, singing his heart out to his new mate who gazed up at her Romeo in rapture from a nearby branch. The bunnies' coats darkened; the coltsfoot defied the still-frozen ground and began to bloom.

Oil patch workers I met on the road that passed the tower would always ask if I had seen bears yet; one fellow told us that a female grizzly with a cub had been seen not 10 kilometres from the cabin. I delighted to hear that there was yet another chance at bolstering their fragile population right in my neighbourhood.

I was still struggling with my responses to Tamarack. The feelings it generated were so very different from the other two towers. I just couldn't seem to find the beauty here. Not that there was ugliness, rather it was more nothingness, a distinct lack of the exceptional. Yet the energy was peaceful, and I felt I could settle here.

The difference in the aura here was reflected in my habits. Where I had never turned on an incandescent light at either Connaught or Luna, I did so routinely now. It just didn't seem to matter if city habits infringed upon us here. Maybe it was the road, suggesting the feeling that this couldn't be wilderness if I could drive my car out anytime I wished; perhaps it was the lack of a definable sense of beauty. I couldn't find the inspiration to decorate the loo. Neither was I moved to put film in the camera; I just didn't seem to see any image I wanted to capture to share with friends. I couldn't be bothered to completely unpack my luggage.

I was able to reconcile a few things, such as any ideas that I could ever make the cabin amount to anything. I accepted that the patient was beyond help, and that, while I would keep it warm and clean, I would not prolong its agony with thoughts that it might ever regain its vigour. Instead, I decided the outside held a shred of hope. If I could at least get it looking neat, much like a quick haircut, it would be half the battle won.

I started to prepare the gardens on those rare days when the ground was thawed. The sod had had a good head start at taking back what had previously been tilled around the cabin and in the larger vegetable garden. I peeled back the grass overgrowing the concrete stoop and gradually uncovered a beautiful stone walkway that someone had obviously taken a great amount of time to create. Perhaps there was a bit of character that could be brought out in this old cabin.

I began making contact with several of the neighbouring tower-folk and found them to be generally friendlier than previous teams, with a good sense of camaraderie and a more relaxed, positive attitude toward the work at hand.

One evening's conversation was particularly poignant, since a towerman who had a lake close to his tower told me about an incident that brought the cold reality of Mother Nature's ways home to him. His tale told of a spring day, the lake still iced over, when a moose chased another down to the lake and out onto the ice where the fleeing animal fell through.

The moose struggled valiantly. Its fight for life went on for many more hours than the towerman would have ever chosen to observe. The ravens came and pecked its eyes out even before death came to end its torment. There was little solace in the fact that the animal was probably in shock long before its final breath.

The following morning the body floated in an ice-free lake. If only that chase had happened a day later, the sacrifice would have been unnecessary, those hours of struggle replaced by a simple swim across the lake. But like a phoenix, the remains of the beast continued through nature's cycle after both a bald and golden eagle as well as the ravens spent many weeks feasting on the carcass.

Thoughts returned to a seminar I attended at a kayaking symposium many years ago on back country safety and first aid. The seminar leader posed a scene where one of a group of paddlers in heavy, stormy seas overturns and fails to re-enter the kayak on his own. What would you do? I am certain heroic thoughts streamed through everyone's mind. But the truth was, you would paddle away. In seas that high, letting another panicked human touch your kayak would be certain disaster for you as well. You would have to paddle away.

Those thoughts in turn led me back to memories of the day Ddreena left us. I had imagined myself so strong and able to set my jaw and cruise through that situation. How wrong I had been.

I thought also of a towerman who had watched as a cougar walked away with his little dog in its jaws; and of a young father who recently had seen his two little girls killed by a falling tree during a hike in the mountains west of Calgary, one dying in his arms.

The modern world has drawn us so far from the reality of Mother Nature's ways. The personal journey that begins with the experience of observing death deepens the perception that life is indeed a complex circle, with the links integral and unbreakable.

Pull Up a Tree, Are You Staying Long?

The flow of life, at times feather-soft, at other times as gritty as sand-paper, conjures the image of a person standing in the middle of a busy street with people and vehicles moving like automatons around her, continuing on their journeys without a backward glance over their shoulders. In my imagination, the person is awake, can feel the movement around her; and she has a smile on her face, as if a fresh breeze has cooled her face on a hot day.

Tamarack gave me this feeling, as I began to share the world with its inhabitants. I'm quite sure they couldn't have cared who was working out in the garden; they had their own places to go, food to find, nests to build. And yet I'm sure that my relationship with them, such as their assessment of whether Ted and I were threatening, was unique. The bunnies and the grouse all felt that there was no need to worry, as did the juncos. All could be found within several steps of the door of the cabin at any given time of the day as they quietly went about their business of grazing.

My thoughts continued in that vein as more and more creatures introduced themselves to us. Would they be staying with us for the summer, or were they just passing through? The first flock of robins disappeared with the new snow, leaving me to suspect that they had moved to a warmer part of the woods. But when a second flock of

40 or so individuals stayed about the same length of time, then disappeared as mysteriously as they arrived, I was left to conclude that Tamarack was merely a stop on their migratory map, a place to rest and feast before finishing the trip to wherever their special nesting place might be. Checking the map in the bird guide confirmed the same about Harris's sparrows, who were also merely stopping by for a brief respite before continuing their journey to the Arctic.

So many species arrived during those May days: horned larks, white-crowned sparrows, white-throated sparrows and black-and-white warblers, some staying and some continuing on their way. Many others like the Swainson's thrushes, Wilson's warblers and yellow-rumped warblers lingered long enough to suggest they would be staying. Others, like the northern waterthrush, evening grosbeaks and rose-breasted grosbeaks were gone in a day. A common snipe made use of our rapidly diminishing frog puddles.

It was turning out to be a long, cold spring. The Victoria Day weekend, the traditional gardening weekend for many parts of Canada, brought more snow, with the temperatures still dipping below freezing at night. If the day was sunny, it would be pleasant enough in the cupola, but if the clouds blocked the rays it wouldn't be long before I'd be scooting down the ladder to wrap my frozen fingers around a mug of hot chocolate. The trees couldn't warm up either. I watched them to report green-up stages with my weather, but they just didn't seem to be able to bring themselves to unwrap those new tender leaves in the face of the persistent cold winds.

It was easy to tell if it was one of those rare warm mornings as soon as I poked my head out from under the covers, because the birds would be singing joyously, whereas on cold mornings the woods would be almost silent.

But to celebrate the long weekend, and in the hope that preparing for spring would quicken its onset, I finished the alder and willow fence around the flower garden in front of the cabin to try to keep

Green Up Stages

<u>CGS (Cured Grass Stage)</u>: 75% or more of the grass in the surrounding area is "cured" (brown/dried).

<u>TGS (Transitional Grass Stage)</u>: 25–75% of the surrounding grass is green.

<u>GGS (Green Grass Stage)</u>: 75% or more of the surrounding grass is green.

<u>DCB (Deciduous Closed Bud)</u>: 75% or more of the aspen buds are closed.

<u>DOB (Deciduous Open Bud)</u>: 25–75% of the aspen buds are swollen to the point where green parts are visible.

<u>DLO (Deciduous Leaf Out)</u>: 75% or more of the aspens have opened their leaves.

<u>DCC (Deciduous Colour Change)</u>: 50% or more of the aspen leaves have changed colour.

<u>DLF (Deciduous Leaf Fall)</u>: 50% or more of the aspen leaves have fallen.

<u>CCB (Coniferous Closed Bud)</u>: 75% of the spruce tree buds have intact sheaths.

<u>COB (Coniferous Open Bud)</u>: 75% or more of the bud sheaths on the spruce trees have fallen off.

<u>CNF (Coniferous Needle Flush)</u>: 75% or more of the spruce trees have extended their new growth at least 2 cm.

the bunnies out. However, once the gardens were underway, I couldn't stop myself. When the next day warmed just that little bit more, I planted the frost-hardy flowers such as the pansies and stocks.

The following morning I rushed to the windows to see if the flowers were all still there. I laughed as I watched Mama and Papa Whisky-jack bring the family right to my front door, the youngsters easily hopping over my fence and pulling flowers off the plants. You just never know who your opponents might be in a game of garden tug-of-war.

My planting seemed to inspire the weather gods, because the next few days were warm, frost free and scattered with showers. The vegetable garden, complete with a wooden rail fence and gate, and wrapped in chicken-wire mesh to keep the critters out, began to take shape.

The soil was wet and heavy but of a better quality than I thought I would find. Scrap lumber piled at the side of the yard became the walkways between the beds, old chicken-wire fencing would also be reclaimed for the snow pea trellis. Lacking a proper hoe, I found the Pulaski axe did the job of breaking up the soil, but unfortunately the short handle had me working at a 90° angle. By day's end I limped back to the cabin and headed for the aspirin bottle. But is there a more wonderful exhaustion than that achieved by physical labour? I think not.

Jupiter chased the waxing half moon across the evening sky; the horizon still held the last light of day at 23:00. It was only a moment after my head hit the pillow that sleep mercifully quietened my aches and pains.

Ted sits on our reclaimed stone sidewalk; the alder and willow fence around the garden keeps the hares out.

Unique Characters All

The sound of rain on the windows, the rain barrels overflowing. Wet bunnies munching fresh green delicacies on the lawn. A hairy terrier who dislikes being wet, lying in front of a warm stove trying to dry his tummy. Hot chocolate by my side, my lap warm from the sweater in progress, now big enough to double as a blanket.

These were the tidings of warm days in the Weasel Hills. Until recently, the only ingredient lacking for a successful spring was the rain, which had finally arrived with a passion. The forest responded joyously, the trees seemingly exploding in vibrant greens of both leaf and catkin. It was about time.

The prelude of warm days before the rain had allowed me to attack the garden with a vengeance. The transplanted rhubarb seemed to be coping, and chives from the old garden site were finding places in the new garden as time allowed. Seeds had been tucked in and were now being cajoled from their sleep by the rains. With space yet remaining, I considered transplanting some wild raspberry canes to see how they might feel about being domesticated. Nothing ventured, no berries gained (although I pondered whether I was just sending out invitations for a late-season bear garden party).

Ted and I took advantage of the low hazard level and took an especially long walk to get to know the area a bit better. This young/regrowth transitional-muskeg area was fascinating, since each grove or rise was an individual. The mature, dense stands of black spruce that appeared to have escaped the wrath of several fires over the years were dark and mysterious, the trees distinctly tall, slender and unattractive, and draped in long strands of moss. Precious little light penetrated to the forest floor. The unique subset of songbirds that calls these stands home was the most difficult to identify, because the density of the trees hid the birds from view and masked where their songs originated.

The mature pine stands whispered with the breezes, carrying a much thicker covering of lush mosses, lichens, Labrador tea and bearberry below them. The silvery cladonia was in high contrast to the bright green mosses; the waxy bog cranberry leaves shone. The pure muskeg areas, with the least coniferous growth, were a combination of woody shrubs, like alder and birch, interspersed with juvenile pine and spruce, all growing from a deep, wet bed of mosses. The roads were edged in alder and willow thickets that housed their own unique subset of songbirds.

In these last days of May there were still no tadpoles in the ponds, but lots of fresh deer and moose prints decorated the soft muddy edges. Frog songs came from the puddles behind the cabin, lulling us to sleep each night.

Whisky-jack and raven youngsters came with their parents most days. While the whisky-jacks would spend more time on the mowed area of the lawn, the ravens tended to the mulched side. It was fascinating to watch them pick apart the mounds of sticks in search of insects, mice or voles. One morning, there was such a racket in the yard that I thought there was something being murdered. I pondered whether I really wanted to look, but curiosity finally won and I glanced out the window. All that noise was coming from one fully grown raven youth who was adamantly insisting his mother continue to feed him. She was doing an excellent job of completely ignoring her child, even though it stood right in front of her with mouth gaping, the most ghastly caterwauling spewing forth. In time he finally gave up, and mimicked his parents' actions instead, choosing his own mound of sticks to deconstruct.

A neighbouring towerman shared the most interesting story one evening. The afternoon had found him watching from his cupola as a marten stalked a rabbit in his yard. The hunt took over two hours before the inevitable end came directly beneath the tower. Not to miss an opportunity, the towerman charged down the ladder, frightened the marten, and won himself a fresh rabbit dinner. I chuckled as I imagined the vegan towerperson just to his west giving a shudder.

That started me thinking about the diverse bunch I had crossed paths with since embarking on this tower journey. It was true what a forestry friend had said: I had yet to find the tie that bound these folk. The retinue included vegans, carnivores, artists, university students, IT professionals, writers, retired armed forces personnel. They were young and old; male and female; husbands and wives and parents. Pets came with some, others were without. Non-communicators contrasted with others who were chatty. There were those who would only view, but never touch, the wildlife around them, and others who enticed wildlife into their yard with treats such as saltlicks; others who harvested what they could.

Notwithstanding all these differences, there was an interconnected respect between us. We were there for each other, and we all took pride in performing our duties. Although we knew each other by voice, we would probably never know what our neighbours looked like.

Differences did exist among district teams, and I enjoyed being back in the north again. Maybe it was because of the sunshine and longer days that they were a cheerier bunch here than in the south.

The yard took as many days to dry as it had rained. The rain had not only made the earth green again, but it had also turned the skies back to that rich Alberta blue and brought the clouds back to life.

The days were warm. The forest greened. The bunnies darkened. The birds sang rich, vibrant songs.

I had a couple of surprises then: an American redstart revealed himself one day, and an unexpected visit from an HAC crew brought me fresh water and a new Stevenson screen. The screen was installed more to my height—there would be no more standing on my toes to read the thermometers.

I asked the crew if they would help me drag an old wood stove I had discovered out of the woods, and they cheerily accepted the challenge. It was a rusted, beaten-up hulk with one side caved in. But I could see potential in it. I could see it all painted up in vivid colours with flowers growing from all the burners and oven. After the guys left, I dragged it over to the spot I had imagined for it, then just sat on the lawn with Ted and smiled at it for a while.

Since arriving, I hadn't been able to find inspiration in this place. And here I was smiling at this sad rusty piece of trash metal on the lawn. The first swallowtail of the year landed on the lawn near us; the blues fluttered all around.

I had been contemplating the pile of postcards that usually adorned my outhouse for weeks and hadn't been able to find a moment when I would have enjoyed hanging them. As I sat staring at this wood stove, I suddenly decided that my bedroom was uglier than the outhouse, and that the postcards should be beautifying those walls instead. An hour later, all the postcards were hung. Pictures that brought thoughts of a variety of friends would be all around me as I dreamed.

I went up the ladder and turned my face to the wind. And I smiled. When I went to bed that night, I had that warm feeling on my face from sun and windburn. It was warm enough to leave the window open over my bed. The night breezes wafted in; the full moon cast its silvery glow over the yard.

Maybe I could have a good season from an ugly cabin. Crazier things have happened.

Whichever Way the Wind Blows

The needlework in front of me was nearly completed. I have always derived a great sense of accomplishment from embroidery; plain fabric brought to life with a spectrum of coloured threads. I began thinking of which project I would start next.

The clouds had crept in and the sky had darkened. After a string of pleasant days, it looked as if it was going to actually rain. I still wasn't fine-tuned to Tamarack's weather patterns. The last few afternoons herds of cumulus clouds had accumulated that hadn't managed to accomplish anything precipitous. I chided myself to be patient: when Mother Nature was ready to deliver, she would, and not before.

Accomplishing tasks in synchronicity with the weather, not when it necessarily suited me, was an intriguing step backward from the frenetic modern lifestyle. That thought had come back to me as I prepared to leave for Tamarack and a friend and I were discussing bathing. I had shared the incomplete thought that bathing sometimes was directed by the availability of water. She suggested it would be better to set showering on an every-second-or-third-day basis, with sponge baths in between. But, I replied, it's not easy to set such a schedule since that day might end up being cold or windy. And who wants to stand in their backyard and shower in those conditions? In fact, sponge baths suffice on all days in which you look out the cabin door and shiver with the thought of standing unclothed in the current weather conditions. Lack of available water can also be the deciding factor. Mother Nature is often who makes the decisions at the towers, not I.

All sorts of chores at a tower site are directed by the weather. Quite simply, if the hazard level is extreme, many things don't get accomplished on the ground. The grass gets long. Motivation to cook complicated meals dwindles by the time you look through the cupboards at 20:30 in the evening (later if you've gone for a walk after being cooped up in a small space all day), and the pile of laundry grows larger. This kind of "letting go" could cause eyes to bug in this modern world where city schedules are often so tight.

The alarm clock is one urban discipline that carries over. In fact, it takes on a greater significance, given that there are no days off. The alarm sounds each and every morning, whatever time is best to have weather ready to read around 07:30–08:00 hours. That same regimen extends to afternoon weather and scheds.

But after that, the discipline is to learn to let go. Each day is taken as it comes, to use in the best way possible given its individuality. I remember frustrating moments in my first year as I let go of old routines; now I embrace the adventure of deciding the day over breakfast.

If the bugs are bad in the evening, gardening gets done in the cool of the morning. If the hazard level allows me part of the day on the ground, and the lawn is dry, then I can get the lawn mowed. On rainy days, I bake and do a bit of meal planning, or I can curl up with a good book. On warm breezy days with a bit of morning time on the ground, I can get laundry done. By the end of each season, hindsight proves that the right days were doled out in the right amounts.

If I might borrow from a Julie Andrews quotation, "For me, it is a kind of order that sets me free to fly." I have no long-term schedules to adhere to, therefore I race no longer, and yet, there isn't a day that's not full of activity.

So the embroidery that was only a day away from completion remained on the frame in the cupola for a week longer as rainy days continued. Once, my comfort zone had been so small that I would have been in a dither about such things. Now the notion of deadlines (except tower closings) passes by as if on a breeze, and the ceremony of removing that project from the frame was put on hold while Ted and I went on long, cold, wet walks down the road. We photographed the first spring flowers, listened to the wolves sing and came home to the aromas of fresh baking that greeted us outside the cabin door. I spent the final low hazard day before returning to the cupola getting as many ground chores done as possible. The laundry was done, Ted was bathed, gardens were weeded and a bit of painting was completed.

I defer the scheduling to nature easily now, and that trust has brought the joy of spontaneity back into my life.

The Circle of Life

It was federal election time, and I was determined to understand all the bafflegab so that I could make an intelligent vote. I had been listening to the radio too much during those days. The more I listened, the more disheartened I became. I had come to the point where I feared for the future of Canada.

At the same time, the policies of the current U.S. administration continued to wreak havoc on the world. Bombings, beheadings, closed borders, supposedly honourable nations thinking themselves above observing basic human rights like the Geneva Convention. I also feared for the world. There was so much time to ponder it all.

I was grateful for my reclusive space in the Weasel Hills, where none of the denizens of the forest much cared about what depravities humans were wreaking upon themselves.

One midday, as I went out into the yard, I heard the wolves singing. There were many voices, and they at least, unlike humans, seemed to be able to join their voices that day in harmony.

The very next day, a moose walked up the driveway. It would repeatedly stop and retreat into the woods, then peek up the driveway before stepping out again. I couldn't figure out what was startling her, until I realized it was the flag at the edge of the yard flapping in the wind. The following day I watched for part of the afternoon as she quietly browsed the east side of the site. I found it amazing how a beast that size could disappear behind a small conifer. Between shade and coloration, she was very effective at blending in.

Rain arrived, then thunderstorms, the accumulated precipitation allowing me time in the cabin. As the weather cleared, Ted and I went out for a walk once the road had dried a bit, taking advantage of the chance to walk farther down the road than we usually did. Ted lagged behind as usual, but as we neared the point where I thought we might turn around, he ran ahead. At the same time I began to smell some sort of large, wet ungulate. I checked the woods, but could see nothing. I walked on another few steps, detected the same odour, but still couldn't see or hear anything. Right about then, Ted found something interesting at the edge of the road.

I walked over to find a fairly complete set of intestines, obviously from an ungulate. Wolf tracks and scat were everywhere. I thought of the howling that had come from this general direction. Pondering the scene further, I was amazed at how clean it was: neatly removed intestines, a lack of other organs, no hooves or skull, just a few bits of hair. I couldn't believe that wolves would have been bold enough to take a grown ungulate (although it could have been sick or injured), and that they would have cleaned up every last bit. As I looked around yet again for more clues, I noticed the knife. It had actually been either poachers or Aboriginal hunters.

Thinking that Ted and I should count ourselves lucky and be on our way before a wandering bear also caught a waft of this fragrant cache, I pocketed a tuft of hair and we were on our way. I called my supervisor when we got back to the cabin, and he said he'd pass along the information to the Fish and Wildlife office.

For the same reasons I disliked the hunters coming around Luna, I was discouraged that irresponsible people would leave such a bear-attracting mess so close to my cabin. I shared my story with Judy, and she related a story about a fisherman who had gutted a day's catch on a picnic table not far from her cabin, wandering off and leaving her to clean up the mess and bleach the table.

A full seven days from the wolves' concert, I looked down from the cupola and watched as Mama Moose walked through the yard, her new calf at her side. What a grand little fellow it was! Happy and healthy, so gangly on those long legs, bright red in colour compared to his mother's mottled dark-ochre brown and grey coat. It was almost as though she'd brought him around deliberately, to point out that the cycle inevitably starts again, that as some individuals exit, others enter, and all remains in balance.

As I watched the pair gracefully pick their way through the underbrush, something else caught my eye. A mature red-tailed hawk landed in a tree just behind the cabin, looking in my direction as I peered at him through the binoculars. An unfamiliar cry caught my ear and I looked around again to see an American kestrel land on the top of a tall dead spruce just beyond the west edge of the yard.

For whatever might be happening in other parts of the world, at this moment things were status quo in the Weasel Hills. Mama and Baby Moose meandered through the woods; the raptors continued their hunts. The sun was sinking in the west; the watch was over; dinner was waiting. Ted did his greeting dance at the bottom of the ladder and got a big hug for making me feel so special. The woes of the world fled into the distance again.

Mama and Elliott, so named for a moose from one of my young nephew's storybooks (and easily changed to Eleanor if I ever managed to figure out the answer to that question) stayed in our general vicinity for the next few days. While Mama was slow and lumbering, Elliott was a typical child, gay and full of energy. Once I was in my cupola, Mama felt comfortable to wander about the open spaces on the east side of the site while Elliott preferred the secrecy of the trees.

While Elliott played peek-a-boo, Mama trundled over to one of the frog puddles for a drink. Turning about and realizing that Mama was farther away than what he found comfortable, Elliott galloped over to her and jumped into the puddle. Having quenched her thirst, Mama exited and resumed her browsing. Elliott, however, was fascinated. He splashed about and pawed the water, sniffing and investigating the edges and floating bits. I watched as they gradually grazed their way over to a small grove of conifers and lay down for their mid-afternoon siesta.

Elliott joins his mom in one of the puddles in the yard.

Just around dinnertime a large helicopter landed in the yard, coming in at a low angle right over where the moose had been relaxing. I wondered what Mama would think of that. If she really was the same cow that had been here in previous years, I hoped she would be used to such noisy and windy interruptions.

Two days later I caught a glimpse of Mama in a small clearing hidden behind a grove of pines at the edge of the yard. I was delighted that they had decided to stay. But that day also brought the road grader and two pickup trucks up the driveway, and later I watched as Mama and Elliott came out of the woods and headed toward the road. I lost sight of them as they turned the corner.

I hoped they'd be back.

Watch What You Ask For!

The topic was energies, as Judy and I were chatting one evening on the phone. The belief that there are energies that affect each individual in their surroundings is a personal one. I no doubt denied them in my youth as I dashed about and worked hard on my career, but now with more years and experience, I had found that they were an ever-constant companion in my travels. They often had wise things to tell me, if I could just be quiet enough to listen.

I was describing to Judy how Connaught had enormous amounts of positive energy, an intense sense of euphoria. For as beautiful as Luna was visually, it was rife with negative energy that had frazzled me the entire season, and I found I didn't like myself there. Tamarack was different again; it was as if it had no energy, or perhaps just a quiet energy. It was difficult to describe since it had its positive effects. Here it was like being cradled, although the maternal figure doing the cradling wasn't humming, she was quite silent. I had slept peacefully since arriving; I was calm and satisfied, but not necessarily exuberant as I had been at Connaught. The days flowed past in a quiet ambience like leaves on a meandering stream, and while pleasant enough, they weren't exciting. There just wasn't much to write home about.

As June came to an end, the weather warmed and rain remained elusive. The hazard level crept up to high, and the long days in the cupola began. Fire crews started day-basing at points around the district, including Tamarack. The first few days brought a succession of crews aboard a huge bright yellow 212 helicopter. It was amazing to see a beast that size land in the yard, but each time the ensuing tornado wreaked havoc. It blew the garden gate off its hinges, blew shingles off the sauna and rolled empty fuel drums off into the forest.

Although I initially lamented the additional burden to my budget, I felt responsible to help keep up the lovely little sauna that had been born of a previous towerperson's generosity. So I ordered all the supplies, and pondered my first solo roofing project.

Although the roof couldn't have been much simpler, each slope being three metres by two metres, it did have a stovepipe that would need to be negotiated. I lamented being short in stature with short arms *and* having a stepladder that was too short *and* a roof too steep to walk on. How that huge piece of roll shingle managed to come off given the number of nails applied was a mystery unto itself. It took the better part of a day to peel the old shingles off, pull all the nails and scrape away the bits around the edges that had been cemented down.

With the first sheet of new shingle came the first lesson: don't apply the black, gooey tar-based cement to the roof and then lay down that nine-foot length of shingle. That only made the phone ring, causing me to lose my grip on the shingle, allowing it to come sliding down towards me, which led to the experiment on what solvent might possibly remove all that goo from both me and my clothing (Varsol did the trick).

Lesson learned, I changed the procedure and tacked down the strip except for the edges, then applied the cement, then finished nailing it down. The first slope was easy, as I could apply the lower strip with the help of the stepladder. By nailing down a couple of two-by-fours on the opposite side, I could walk up that side and apply the second strip by leaning over the peak. Thus the lower strip on the second was a breeze, but I hit another puzzle with the upper strip on that last side. I couldn't nail down the two-by-fours to the opposing side without making holes in the new shingle. So after pondering the matter while completing a few circuits around the sauna, the idea came to me to nail those two-by-fours to the fascia boards. That accomplished, I was able to climb up onto the peak and nail the last piece down from above.

At that point I crawled back to the cabin, directly to the aspirin bottle and muscle rub. I couldn't find a part of me that didn't hurt as I got into bed that night, the last puzzle to solve before I drifted off. Every ache was a medal, however. I was amazed that I could move freely enough the next day to get the cap on and finish the job. By lunchtime I was standing by the cabin, staring at the new roof on the sauna. What a fabulous feeling self-sufficiency brings!

My supervisor called, offering three of us towerfolk a helicopter tour of our areas. I jumped at the chance, enjoying a full half hour in the air. It was great to look over the ridges and see what was behind them, and get a better overview of the topography. Equally as interesting as the landscape was the pilot that day. I had heard her voice on the radio for several days; you don't hear too many female pilots, and she had a very interesting accent. I heard her check in and out as they visited the other two towers. I waited anxiously to meet her. At about five foot three, dressed in t-shirt, corduroys, socks and Tevas, with short grey hair, she definitely wasn't your average pilot. A Swiss francophone, Jeanette had been flying for more than 30 years. Oh, how I wished for an afternoon that we could sit and chat over a cup of coffee! The stories that she would have been able to tell! She was just the sweetest, most laid-back person. I laughed at the note taped to the instrument panel stating that the pilot had to be a minimum of 170 pounds. I somehow doubted that. She winked and told me that when she was alone in the helicopter, she threw some extra weight in the front with her.

The following day brought yet another first: two helicopters in my yard at the same time. The second arrival brought a film crew with a Chilean pilot and a teacher from the school in Hinton; they were storm chasing and taking film footage of crews working on fires. They ended up spending two afternoons enjoying the shade and picnic table at Tamarack. For all the long hours I had been spending in the cupola those days, it was nice to see Ted have some company down there.

Over those days the skies changed and the towerfolk began muttering about lightning. Morning skies were laced with high wispy clouds. In late morning the cottonball cumuli would start to roll in, and by late afternoon they would start to tower. Mother Nature teased, as those first few potentially threatening afternoons just rolled by without a rumble. The towers south of the lake called first strikes toward the end of the week, but the skies around Tamarack remained quiet.

On the last day of June, the lightning arrived, and smoke messages were radioed in as fires began just north of Ursa Lake. I had been watching two storm cells for hours. One hung on the ridge west of me, the other lay directly to my north. It was a breezy enough day, and yet both cells remained motionless. A fire crew landed in the morning, joined by the helicopter with the film crew soon after. The video folks concurred that my storm clouds held potential, but flew off once the Ursa fires started.

Boot Hill Tower called in a smoke just over the ridge to my west a few minutes before it crested that rise and came into sight. Weasel Ridge began calling in detection messages next. I crossed my fingers that I wouldn't have starts as the radios chattered energetically with helicopters being sent in all directions and crews being dropped at fires. The smoke to the west turned rose-grey in the setting sun; I watched as the bombers, silhouetted and monotone in that same rose grey, skimmed the ridge and begin attacking it.

My luck didn't hold. A smoke came up almost due east; then two more came up beyond the edge of the hill to the north. While attempting to keep the information on all three messages straight, trying to find an appropriate moment to break in on the radio (I only ever succeeded in sending in short pre-smoke messages that contained bearing, distance and estimated location—it was all that was needed), and taking down cross shots from neighbouring towers, I could hear the radio room informing the helicopters struggling to keep up on the initial assessments that resources were being stretched. They would just have to wait in line.

All the storm cells and lightning I had noted so far were to my west, north and northeast. In an attempt to gather myself, I turned around just for a moment to catch my breath and found myself staring at a fire a kilometre to my south, just across the road. There wasn't a cloud around to which I could attribute it. It was one tree, then a few, then the orange flames shot up, and it was rolling in mere minutes. Just as I got finished calling in that message, I turned my head to find a smoke rising just past the edge of the firebase to the east, not a 10-minute walk down the road from me. It was the same scenario, taking only moments before it took hold and grew quickly. I called that one in, telling them basically just to put down the location of the firebase, not needing to take the time to work out a location. I listened to the radios as they endeavoured to move aircraft and crews, dispatch equipment (hoses, pumps, etc.), and prioritize smokes. My fires grew. I radioed in again and suggested that if they couldn't get someone over to the firebase soon it might get burned over, including the comms tower and fuel drums stored there (if I had considered the direction of the wind instead of only the proximity to the firebase, I would have realized that this was not a big concern). It wasn't long before a helicopter arrived (finding yet another smoke just east of me on his way in) and wrote up an initial assessment. As fast as they accomplished that (you lose track of time by this point), the bombers arrived.

The air around me was suddenly filled with bird dogs, bombers, helicopters and crews being dropped, like bees over a hive. I had been taking photographs of the fire across the road, and as the first bomber arrived I imagined the amazing shots I was going to get. I rolled the film forward only to discover there were no more pictures left on the roll. The images would have to be consigned to memory, since I couldn't leave the cupola. It was fascinating to watch the bird dog and the bombers disappear into the solid columns of black smoke, flying blind for those few seconds before emerging again, just to bank and come in for another pass. Empty tankers would leave in the direction of Long Lake just as a loaded one would appear on the horizon to take its place. Once there were enough drops to knock the flames down, they were off to the next location.

Whitefish Pond Tower tried to call in new detection messages but couldn't even get the radio room to respond—they were that busy. Tiptop Tower was next to start reporting; out-of-district crews began to make their way up to that northeast area. Helicopters were in and out of the firebase next to me to refuel. One helicopter landed at Tamarack and actually used a half drum that was on-site when I arrived in the spring; I worried about them for the rest of the night. Another helicopter arrived at the fuel cache near Lloyd Tower to find the cupboard bare, and the pilot radioed that he was following the highway back to Long Lake.

At the beginning of that day, fires were being numbered in the high 120s. The last numbers I heard being given out that night were past 150.

I could only imagine when the persons in charge of deployment started work that next morning, trying to assign aircraft, manpower and supplies to that many fires (and still have enough left available to handle new fires should they arise). My tower supervisor became an incident commander and spent the day at the firebase beside me getting the fires around Tamarack under control. On my first trip up the tower at 08:00 to begin searching for hold-over fires, I watched a convoy of pickup trucks and tractor trailers carrying cats (bulldozers) pass on their way to the fire over the western ridge.

More helicopters arrived and the bombers returned, dropping more retardant. Again the air around my tower was alive with air-craft: three bird dogs, assorted tankers with water and retardant,

Only about one kilometre from the tower, this fire grew quickly from the first wisp of smoke I spotted 10 minutes prior to taking this picture.

and helicopters long-lining water buckets. In the afternoon I could hear the sound of cats and chainsaws rising over the trees. As for the fires on the north and east side of my ridge, the visibility was so low that I couldn't see those locations. The chatter over the radios was still relentless, and with so many fire numbers on the go, it was next to impossible to know which fire was being discussed.

Late that afternoon 10 millimetres of rain fell, yet the three towers around me were left dry. Another 11.2 millimetres fell the next day, with only the tower west of me receiving any of that.

It took less than 24 hours for someone to beat me with a fire closer to their tower. Elk Hill Tower, to my southwest, called in a burning tree only 200 metres from the cabin, and yet another start a half kilometre away.

I wondered where Elliott and his mom were and what they made of all the smoke, the fires and all the traffic, both in the air and on the road. I had no doubt that they were well as moose can move quickly, but I wondered in which direction they had run and whether I had seen the last of them.

It is the most peculiar feeling once you've called in your smokes, especially on busy nights such as this, to then sit still. You want to help further. As the first link in the chain, your part is over so quickly, and everyone else seems to be so busy. There is an incredible urge to get involved on the radios which is actually no help at all. I always seem to have this odd itch to start making sandwiches. But a towerperson's assignment is a simple one: stay off the radios except for pertinent information and just keep watching.

By the end of Canada Day, fire numbers were over the 160 mark; by the end of the next day they were past 170, with 25 fires actively burning. Alberta had a total of 37. I listened as the news on the AM-FM radio stated that the total area of the Yukon that had already burned this season was 10 times greater than the total burnt area for the previous year. British Columbia wasn't faring much better. No wonder Quebec was sending tankers and crews to the west.

As the helicopters came and went from the fire base, another motorcade of tractor trailers loaded with bunkhouse trailers began to arrive at the fire base. I counted at least six. Command shacks and a truckload of porta-potties arrived, and a staging camp was born. Ted and I walked down to take a better look that evening. With the tractor trailers carrying the shacks wallowing around in the mud, the crews staying there had formed a tent city for that first night. That morning the site had been nothing more than a meadow; now it was a town.

I fell asleep that night to the sound of trucks struggling with their loads in the mud. The next morning I awoke to the sound of helicopters powering up. With all the expense of setting up a camp of this size, I suspected I would have neighbours for a while. During our morning walk, Ted and I found many deer tracks on the road, all travelling away from the camp.

The radio room first came on the air at around 05:45, and shut down for the night around 23:00. Many of the incident commanders could be heard aboard those first helicopters taking off in the morning; their voices could still be heard on the radios with the last light of day.

The rains that followed that hectic night allowed me a few more hours on the ground for the first time in almost two weeks, but at the same time these post-lightning conditions dictated diligence as I watched for hold-over fires. I was never far from my tower.

But all around me, the forests of the Weasel Hills went about their business of summer. Wild roses and yarrow began to bloom, and the birds were in the air again, indicating that their eggs had hatched and young needed to be fed. Pine pollen had been on the breezes for more than a week, once again coating the rain barrels with pale yellow muck.

Mama Grouse brought the youngsters out onto the lawn early one morning. It was a lovely quiet moment as the little ones bounced along, Mom herding them from behind. Little gifts such as these are always such a pleasure. They brought balance to the long hours that week.

Seven days later the camp down the road was closed after two days of rain totalling over 80 millimetres soaked the Weasel Hills. The whitetail stag returned to graze in the evenings, and mice, not being stupid creatures, decided that my cabin was a better place to stay than out in the yard doing the backstroke. Some other critter decided that under the cabin was a good place to relocate and that it would be fun each evening to snip the flowers off my pansies. Ted heaved deep sighs from the boredom of being cabin-bound; I worked toward completing some knitting projects. The forest around us was quiet again.

Flight Formation

I looked up from my embroidery and watched the first of the flocks of geese fly over Tamarack, heading south. Oh, I just wasn't ready to accept that idea yet.

One of the white birches in the yard had decided to turn blond. Asters bloomed at the edges of the yard. I had never had a yard burst forth in such an abundance of mushrooms, large and small, orange, white and cream. The bog cranberries ripened and turned scarlet; the amount of blueberries in my refrigerator was only limited by the time I could be away from the tower and how long I could bend over to pick them.

The song of the wind through the aspens was changing, and the leaves were just that little bit more brittle, varying the tone ever so slightly.

Yes, the end of the tower season was once again looming dead ahead. I wondered if I would ever get used to that idea. For all the things that had been so wrong at Luna, the length of season had been perfect, from snow to snow. Not that I wasn't aware when I accepted Tamarack that it was scheduled to close at the end of August, but all I could do was cross my fingers and hope for a long, hot, dry season. With only a week left, the days had become progressively cooler, the thunderstorms had rolled in, and the rain began to fall. The afternoon weather report had forecast two days of precipitation. Dammit. I hadn't seen an aurora yet; hadn't sat out and pondered the sky.

I continued to hope for the opportunity of sub work at another tower, but as the time ticked away, the chances seemed slim.

I thought about the young, beautiful jet-black bear that had visited us earlier in the week. He just didn't seem fat enough yet—the end of August must have caught him off-guard as well. I looked around at my gardens. The sweet peas in the flower garden were just getting ready to open their first buds; I had just managed to enjoy my first snow peas. The nasturtiums were at their best; the geraniums were crowned in blooms. It was going to be hard to turn my back and walk away from them. I had, at least, succeeded in my quest to make the yard neat and tidy. But such is the reality of the tower; it really isn't yours to enjoy at your leisure. When the fire hazard is gone, so are you.

The cabin still smelled; I still hated the curtains; the ants continued to march across the floor; the kitchen counter was still lined with mousetraps and the ceiling in the bedroom was still that nasty shade of nicotine mixed with the yellowing caused by the propane heater and stove. But the floor itself had improved with a couple of coats of paint. I had never tried to paint linoleum before, but it had worked, even if the floor was now standard floor-paint grey. At least it could look clean.

It had been suggested to me that I should consider staying at a tower for more than one season; that I was getting the reputation of being a bit of a "tower gypsy." I found it a strange quirk of fate that Tamarack should end up being where I would change my wandering ways. Although we were good friends, we weren't in love. I had been right about the sleepy aura here. While it was somehow comforting, it certainly wasn't exciting. But at the same time the team here was the best I had ever worked with, and I had been treated far better than in any of the previous districts. Life always seems to present the most interesting set of choices.

All the disquieting thoughts about the impending upheaval that filled my head were quelled whenever Ted and I set out for the blueberry patch. Especially the one that we had to make our way through the bush to reach; there we seemed able to get away from everything. The repetitive nature of picking seemed soothing, keeping my mind on being efficient and cleaning all of the ripe berries from a group of bushes before moving on. Once off the road and through the alder scrub, we would disappear into the aspen stand. The ground was littered with deadfall, with some trunks torn apart by bears in search of ants. The south edge of the stand belonged to the beaver that occupied the adjacent pond, but the northern edge contained our hidden blueberry and cranberry patch. Ted would curl up in the mossy, damp undergrowth at the edge while I got on with my picking. And the fruits of the pickings were a wonderful treat in the baking and on my cereal in the mornings.

Another wonderful distraction came with a visit from a good friend, the first friend that had ever visited me at a tower site. It was interesting to see her equate the images she had formed in her mind from my descriptions with what she was seeing now. Ted was thrilled to have a friend to keep him company while I was in the cupola, and I was thrilled to have someone to entertain and to take my mind off the impending closing. I introduced her to the Weasel Hills ecosystem, and we picked blueberries together and made jam.

With her departure, it was time to pack. I cleaned the cabin, filled boxes and packed the car. There was no need to wait for a helicopter this year. My supervisor arrived and we dutifully went through the closing checklist and locked everything up. I was on my way.

"*This is XMG 746, down for the season.*"

**Tamarack Tower
Incomplete Species List
76-02-W5M
Elevation approximately 940 metres
April 23–August 31, 2005**

Aspen

Asters
Asteraceae spp.,
including smooth blue aster
A. laevis

Bicknell's geranium
Geranium bicknellii

Black spruce

Bog cranberry,
Vaccinium vitis-idaea

Bog laurel
Kalmia polifolia

Bog violet
Viola nephrophylla

Bunchberry
Cornus canadensis

Common blueberry
Vaccinium myrtilloides

Common red paintbrush
Castilleja miniata

Common yarrow
Achillea millefolium

Crowberry
Empetrum nigrum

Felwort
Gentianella amarella

Wild mint
Mentha arvensis

Field horsetail
Equisetum arvense

Fireweed
Epilobium angustifolium

Graceful cinquefoil
Potentilla gracilis

Grasses and sedges

Great bulrush
Scirpus acutus

Hooded ladies'-tresses
Spiranthes romanzoffiana

Labrador tea
Ledum latifolium

Lodgepole pine

Marsh marigold
Caltha palustris

Mosses and lichens

Mouse-ear chickweed
Cerastium arvense

Narrow-leaved hawkweed
Hieracium umbellatum

Northern grass-of-Parnassus
Parnassia palustris

Northern or wweet-scented bedstraw
Galium boreale or triflorum

Ox-eye daisy
Leucanthemum vulgare

Palmate-leaved coltsfoot
Petasites palmatus

Pineappleweed
Matricaria discoidea

Pink corydalis
Corydalis sempervirens

Red clover
Trifolium pratense

Red currant

Reindeer moss
Cladonia spp.

Showy everlasting
Antennaria pulcherrima

Stiff club moss
Lycopodium annotinum

Tall larkspur
Delphinium glaucum

Tall smooth goldenrod
Solidago gigantea

Twinflower
Linnea borealis

Western jewelweed
Impatiens noli-tangere

White baneberry
Actaea pachypoda

White birch

White clover
Trifolium repens

White spruce

Wild chives
Allium schoenoprasum

Wild raspberry
Rubus spp.

Wild rose
Rosa arvensis

Wild sarsaparilla
Aralia nudicaulis

Wild strawberry
Fragaria vesca

Willow

Wintergreen
Pyrola spp.

Yellow rattle
Rhinanthus minor

Yellow sweet-clover
Melilotus officinalis

I Am a Fire-Tower Observer

I went to the woods because I wished to live deliberately, to front only the essential facts of life, and see if I could not learn what it had to teach, and not, when I came to die, discover that I had not lived.
—Henry David Thoreau, *Walden* (1854)

Three seasons are now behind me. Each and every tower, each and every season has had its own individuality and I feel honoured to have known each of them.

I have pondered the diverse isolation thoughts of wonderful Canadian writers such as Gilean Douglas, Ken Stange, Grey Owl via the writings of Lovat Dickson (Grey Owl's publisher) and the American wilderness defender Mardy Murie, and have come to realize more clearly that each person's connectedness—their relationship—with the forest is unique. I once pointed out to a fellow towerperson the different paths that led us both to our fire towers: "You go to your tower to be alone, I come here to be with the forest." The depth of that sort of individuality has become so much clearer with experience. One towerperson's experience cannot be compared to others. We all seek our individual solitudes.

Compared with Ms. Douglas' isolation experiences, I have only dabbled in solitude. I always have neighbouring towers to chat with; the radios babble away as the radio room talks with the aircraft; fire crews and helicopters fall from the sky into my backyard for a day now and again. Visitors arrive unexpectedly. Humans never seem far away. Where some towerfolk revel in visitors, I can only partake of small doses before I feel robbed of my space.

I don't believe that working at a fire tower has the ability to bring me any nearer to true solitude than it already has during these first three seasons. The radio will continue to key up each morning requesting weather reports, each local towerperson's voice filling my cabin with their analyses of the weather from their corner of the forest. Helicopters will continue to land in the yard; visitors will continue to arrive.

But many more opportunities lie ahead in my experiments and interrelationship with the boreal forests, which I am both inspired and excited to investigate. As I become more comfortable with the routines, I will find the time to take on new adventures. I need to find a way of describing, painting, or clarifying the intricacies of our forest ecosystems to add to the ongoing struggle to make the modern world take on a more reverential attitude toward what can never be replaced once it is used up in greed. It is the least that I can do for a living breathing entity that has taken me into its arms in such a quiet, loving way and given so much to my life.

I find myself drawn away from the city way of life. I chuckle when friends decline invitations to the tower because there is no power or running water, because I cannot fathom why they feel they need it. The demands of modern living, coupled with the systems necessary when we concentrate humans into cities, bring on this crush of appliances, specialized tools and utilities to fulfill the silliest little chores, when doing them manually often takes less time. Instead of imagining myself in some fancy new inner-city condo, I see myself planning my retirement years in a small cabin or trailer in a rural or wilderness setting.

Although I appreciate being back with friends again and enjoying the arts, culture and cuisine of the city, once November rolls around I'm ready to go back out to my tower. The glow has fallen away from city life; I long for the simplicity of the forest.

Just as my wilderness experiences were a continuing adventure back in the east, so towering in Alberta will be further chapters within a larger matrix of adventures. I look forward to improving my abilities to produce my own food, wildcrafting foodstuffs from the forest and tackling bigger projects around the cabin. I am eager to add more images to my mind of the subtle year-to-year changes as the world continues its turn through the seasons and to hone my skills at hearing and interpreting what the forest says about itself. Someday I will not only realize that the chickadee's song has changed, but I will know what it means. I want to be able to translate the subtle nuances of nature.

It is winter now. That's the only drawback to being a fire-tower observer, you just can't find year-round work. So it's back in the land of electricity, overwhelming technology, of humans and vehicles bustling down asphalt and concrete roads. For as much as I need my winter city life to keep me connected, it drains energy from me. The beauty and the solitude of the forest replenishes me. As with everything in life, there is a balance.

But the sun is now rising earlier each day, and I long to be sitting in my chair, knitting in hand, perhaps looking up to find the white-tail stag peacefully grazing in my yard, or perhaps listening to the chickadees sing in the spruce trees outside my window, finding a new flower smiling up at me, feeling a fresh breeze on my face.

I miss the adrenaline rush when I look out the cupola window and see that small puff of smoke rising above the trees and listen to the ensuing chatter on the radio as fire crews and helicopters are dispatched. I miss the joy of watching the harriers cruise the cutlines, or revelling in the feeling of accomplishment when I put down my needle and cover my needlework at the end of the day.

I've learned what a small footprint I leave from a simple solitary existence. I can take all the blueberries and cranberries that I need and barely put a dent in what is available, leaving more than enough for the creatures of the forest. Four three-metre rows of lettuce allow me more than I'll ever eat. Although not the prettiest of thoughts, one latrine hole lasts for several years and doesn't go through five gallons of treated water with every use. The amount of reading I accomplish each year grows appreciably, allowing so many new and wonderful ideas to fly around in my head, and I have the time to ponder them all. My identification skills include many more birds and plants. I have re-evaluated how long a moment, an hour, a day is (or what it can hold). One of my poorest skills, listening, is starting to improve again (noting that the ability to listen is directly connected to patience, which I might never master).

This coming season may be the first time that I return to the same tower. In one way it will be nice to be able to plan in advance, because I know what needs to be done there. I find myself wondering if Elliott will be back for a visit. But I also lament that I will not enjoy the excitement that a new site brings with the ensuing joys of investigating yet another forest, coupled with the amazement that comes with the discovery of all the tiny ways that forest is unique from all the others.

It is my greatest hope that I will continue to live and work at the towers for years to come. When I finally cannot climb a 30-metre tower any longer, may there be a 20-metre tower that welcomes me. When I cannot make that climb any longer, may a mountaintop tower invite me to its door.

Once again I am counting the days until spring.

And in a place of dark primeval glory
The Old Ones stand and guard the land where they began
    a millennium ago
A poet's dream, a painter's inspiration
The joy of life on every side, as nature meant it so
Oh what could be more beautiful
What could touch my heart
What place on earth could keep me from this part I call home?
— from "Island Home," by Eileen McGann,
on her album *Beyond the Storm*, 2001,
Dragonwing Music, Mill Bay, B.C.
www.eileenmcgann.com

For each evening by the campfire, some adventure to live over again. And each little event, each lovely vista, every wild creature, held importance for us. Our appreciation was keen, not diverted by other people, newspapers, radio, or too many contacts. Have you ever noticed that books read in the wilderness stay with you a long time? Their entry into your mind is unimpeded.

There may be people who feel no need for nature. They are fortunate, perhaps. But for those of us who feel otherwise, who feel something is missing unless we can hike across land disturbed only by our footsteps or see creatures roaming freely as they have always done, we are sure there should be wilderness. Species other than man have rights, too. Having finished all the requisites of our proud, materialistic civilization, our neon-lit society, does nature, which is the basis for our existence, have the right to live on? Do we have enough reverence for life to concede to wilderness this right?

—Margaret E. Murie

Suggested Reading

Douglas, Gilean. *River for My Sidewalk*. Victoria: Sono Nis Press, 1984; and *The Protected Place*. Sidney, B.C., 1979.

Stange, Ken. *Bushed*. Toronto: York Publishing & Printing Co., 1979.

Dickson, Lovat. *Wilderness Man—The Strange Story of Grey Owl*. Toronto: Macmillan of Canada, 1973.

Murie, Margaret E. *Two in the Far North*. Portland: Alaska Northwest Books, 1962.

Born and raised in Niagara Falls, Ontario, Sharon Stratton has pursued many interests, including veterinary technology and advertising sales. After earning an Honours B.Sc. (Biology) from the University of Guelph, she settled in Alberta and discovered tower life. Sharon currently winters in Calgary.